Disaster Response by Ceaușescu's Communist Regime in Romania

This book contains the first comprehensive histo[…]ng extensive primary sources to trace the 1977 earthquake disaster response by the Ceaușescu communist regime, contextualizing its contribution to the public risk that remains in Romania's capital Bucharest. It traces a history of one authoritarian government's disaster response linking its decisions and ultimate inactions to contemporary public risk. The book begins with a stand-alone chapter to introduce readers to twentieth-century Communist Romania and contextualize the Ceaușescu regime's response. It provides insights into how Radio Free Europe filled the information vacuum, how the political police, the Securitate, worked as first responders, and how scientific experts debated the best course of action. It examines how the regime requested specific foreign assistance and activated its Securitate abroad to encourage such, prioritized restoration of the economy, and "encouraged" domestic cash and labor contributions in the name of recovery. The book examines how the disaster response abruptly ended, leaving thousands of structurally unsafe buildings. It explains the contemporary seismic risk and post-communist mitigation efforts to reduce it.

This book will be of interest to students, researchers, and policy-makers in the fields of history, disaster studies, urban planning, politics, and those interested in communist-era Romania, Europe, and Eurasia; totalitarian and authoritarian regimes.

Karin Steinbrueck is a modern European historian and is an assistant professor at National Louis University, Chicago. She has lived and worked in Eastern Europe and the former Soviet Union as a Peace Corps Volunteer, Fulbright Scholar, Boren Fellow, and Foreign Language and Area Studies grant recipient. She lives with her family in Evanston, Illinois, USA.

Routledge Studies in Hazards, Disaster Risk and Climate Change

Series Editor

Ilan Kelman

Professor of Disasters and Health at the Institute for Risk and Disaster Reduction (IRDR) and the Institute for Global Health (IGH), University College London (UCL)

This series provides a forum for original and vibrant research. It offers contributions from each of these communities as well as innovative titles that examine the links between hazards, disasters and climate change, to bring these schools of thought closer together. This series promotes interdisciplinary scholarly work that is empirically and theoretically informed, with titles reflecting the wealth of research being undertaken in these diverse and exciting fields.

The Post-Earthquake City
Disaster and Recovery in Christchurch, New Zealand
Paul Cloke, David Conradson, Eric Pawson and Harvey C. Perkins

Local Adaptation to Climate Change in South India
Challenges and the Future in the Tsunami-hit Coastal Regions
Devendraraj Madhanagopal

Good Practices for Disaster Risk Management of Cultural Heritage
Practices of ITC participants
Edited by Rohit Jigyasu, Kim Dowon and Lata Shakya

Agency in Poverty and War
Consciousness in Rural Human Development
Vivianna Rodriguez Carreon

Religious Responses to Earthquake and Volcanic Eruption Disasters
David Chester, Angus Duncan and John Duncan

Disaster Response by Ceauşescu's Communist Regime in Romania
The 1977 Earthquake
Karin Steinbrueck

For more information about this series, please visit: https://www.routledge.com/Routledge-Studies-in-Hazards-Disaster-Risk-and-Climate-Change/book-series/HDC

Disaster Response by Ceauşescu's Communist Regime in Romania

The 1977 Earthquake

Karin Steinbrueck

Routledge
Taylor & Francis Group
LONDON AND NEW YORK

First published 2025
by Routledge
4 Park Square, Milton Park, Abingdon, Oxon OX14 4RN

and by Routledge
605 Third Avenue, New York, NY 10158

Routledge is an imprint of the Taylor & Francis Group, an informa business

British Library Cataloguing-in-Publication Data
A catalogue record for this book is available from the British Library

ISBN: 978-1-032-63291-9 (hbk)
ISBN: 978-1-032-63292-6 (pbk)
ISBN: 978-1-032-63293-3 (ebk)

DOI: 10.4324/9781032632933

Typeset in Times New Roman
by SPi Technologies India Pvt Ltd (Straive)

Dedicated to all who lived through or lost someone as a result of Romania's 1977 earthquake.

Contents

Figures

Acknowledgments

Thank you to all who helped with this history of the Ceauşescu regime's 1977 earthquake disaster response. Charles King of Georgetown University encouraged me to write about the 1977 earthquake, the topic of my PhD dissertation, advised by Benjamin Frommer and read by committee members John Bushnell and Peter Hayes at Northwestern University. I am thankful to all the staff at Consiliul Naţional pentru Studierea Arhivelor Securităţii (CNSAS), Arhivele Naţionale ale României and its Direcţia Municipiului Bucureşti, Fundaţia Academia Civică, Memorialului Victimelor Comunismului şi al Rezistenţei, Fundaţia Culturală Memoria, and the Hoover Institution Library and Archives at Stanford University. Engineers Emil-Sever Georgescu and Antonios Pomonis shared data, answered countless questions, and collaborated to help bring the scientific understanding and human experience of disaster together. Cristian Capotescu and Luminiţa Gatejel helped as colleagues in disaster response. The Society for Romanian Studies; the Association for Slavic, East European, and Eurasian Studies; the Summer Research Laboratory at the University of Illinois Urbana-Champaign; the Romanian Studies Organization at Indiana University; and the Northwestern University Russian, Eurasian, and East European Studies Research Program, each provided collaborative venues for me to develop my work. Robin Bacsfalvi, Alexandru Budişteanu, Carol Levy, Joel Levy, Laura Raicovich, Romulus Rusan, Ambassador Pamela Hyde Smith, Olga Ursu Stefan, Liviu Tofan, Andrei Ursu, and Richard "Dick" Weist enlightened me with their experience of the 1977 earthquake and 1970s Romania. Special thanks to Elis Neagoe-Pleşa of CNSAS, who encouraged my research into the 1977 earthquake and helped in countless ways along the journey. I am grateful to Doina Angelescu, Ramona Angelescu Naqvi, Vlad Bolocan, Lorenu Tecu, and Ioan Valentin for their invaluable help in Bucharest. Lastly, I am grateful for my family's perpetual support. All errors are mine.

Introduction

Disaster response is helping. Governments activate response protocols. Nongovernmental and international organizations send humanitarian and technical assistance. People not directly affected send care packages, and those directly affected respond as they can. Temporally, a response begins immediately after a disaster and ends either when people are well again and life returns to normal or when those in charge say it is finished. The latter was the case in Romania in 1977, following the 7.4-magnitude earthquake that began at 9:22 p.m. on March 4. The earthquake killed 1,578, injured 11,321, and damaged more than 200,000 homes.[1] Felt in two dozen neighboring countries and all across Romania, it was the second-strongest earthquake to shake the country in the twentieth century. The decisions made during the communist regime's earthquake disaster response under authoritarian leader Nicolae Ceauşescu (1918–89) have implications to this day, with hundreds of structurally unsafe buildings ready to fall in Romania's next major or great earthquake.[2] The Ceauşescu regime engaged in disaster response practically nonstop for weeks and then intensely for months, until July 1977, when it ended assessments of and necessary repairs to damaged buildings, instructing experts to do so and excluding the public from this fact. This is the first comprehensive study of the Ceauşescu regime's disaster response, which included both typical and atypical activities that ran short- and long-term, spanning from the moments after the earthquake struck well into 1978.

The 1977 earthquake struck without warning, but it was not unexpected. Romania sits in one of the most seismically active regions in Europe, with more than a dozen recorded earthquakes with a magnitude above 7.0 on the Richter scale from the sixth through the twentieth centuries. Earthquakes are commonplace, with one low-level quake registering every few days as Romania sits atop the meeting place of three tectonic plates, whose movement can create "intermediate-depth" or "deep" Vrancea earthquakes, named for the county where their usual epicenter sits. While not built on a tectonic fault like Istanbul or San Francisco, Romania's capital, Bucharest, is particularly vulnerable to earthquakes, due to its proximity to the Vrancea epicenter, the city's population density, and large movements just beneath the upper mantle of the earth's crust. By 1977, Romanian scientists were well aware of the seismic risk. During

DOI: 10.4324/9781032632933-1

the communist era, governments passed new building codes that considered seismic resistance, and a month before the earthquake, the state opened the Seismic Risk Center. It had not, however, adequately repaired damage from Romania's previous major earthquake, a 7.7-magnitude quake in 1940, which significantly contributed to the devastating structural damage in 1977. Bucharest experienced the overwhelming majority of this damage and the resulting casualties: 90 percent of the deaths, 67 percent of the injuries, and 90 percent of the industrial and residential damage.

Moments after the 1977 earthquake's vibrations ended, the disaster response began. That experience and the moments that followed open Chapter 2, which narrates what first responders, government officials, and ordinary citizens saw, felt, and did. It traces the first disaster response activities and the Ceauşescu ruling couple's initial impressions of the damage on the morning of March 5, when they returned from abroad. Typical activities had first responders searching for survivors and retrieving victims, fighting fires, and rerouting traffic. Health-care workers treated the injured. Workers cleared away the rubble and debris. In the days that followed, engineers and specialists started assessing the damage and recommending repair plans, and regime officials organized foreign aid receipts. Romanians stepped up and helped, too. They identified the dead, carried and drove the injured to the hospital, shared their homes with the displaced, joined work brigades, and donated cash and in-kind goods. In response to the regime's information vacuum, the Munich-based Radio Free Europe (RFE) immediately ran nonstop and live on-air programming, connecting Romanians inside and outside the country.

The Ceauşescu regime used its political police, the Securitate, throughout the disaster response, an atypical activity. The Securitate went beyond surveillance and information gathering, to working alongside and even in the guise of first responders, guarding worksites, digging through rubble, and deterring looters, also discussed in Chapter 2. Securitate members inventoried, collected, and returned anything of value found and helped steer the regime's foreign assistance preferences, in particular by activating its clandestine network abroad, discussed in Chapter 3. To contextualize the Securitate's role in the disaster response, an overview of its organization, reorganization, and role during Ceauşescu's tenure appears in Chapter 1. Securitate reports most directly inform Chapter 4, which examines how the regime noticed that Romanians were helping and turned those voluntary, benevolent activities into mandates. It left most citizens to make their own home repairs and "suggested" unpaid shifts and wage contributions from workers well into 1978. It exploited, too, Romanians' close family ties, allowing most of those displaced to find shelter with family and friends.

Today a serious public safety risk exists in Bucharest from the buildings damaged by the 1977 earthquake and left unrepaired and structurally unsafe by the Ceauşescu regime. Almost four hundred multistory buildings are likely to collapse, and hundreds of others are expected to experience significant, injury-causing damage when the next major or stronger earthquake occurs.

That public risk is a direct legacy of the Ceauşescu regime's disaster response, which had two distinct phases. The first took place during the first four months following the earthquake, when, as discussed in Chapter 5, experts undertook a robust assessment and repair recommendation process for tens of thousands of damaged structures while regime officials reviewed reports made after Bucharest's 1940 earthquake to learn why so many buildings fully or partially collapsed in 1977. Experts and regime officials debated whether demolition or repair was the best course, with Ceauşescu initially firmly on the side of repair. He gave two deadlines during the disaster response. The first, established early in the disaster response for May 1, was the deadline for experts to stop making assessments and reduce the number of evacuations from damaged buildings. The second phase began on July 4, when, as traced in Chapter 6, Ceauşescu ordered the end of all structural assessments and a hastening of repairs to the damaged buildings, with work to be completed by August 23. This second deadline marked the end of the regime's disaster response. This July cessation order is the primary reason why thousands of buildings in Bucharest today are still structurally unsafe. Chapter 7 describes the risk mitigation efforts taken by several post-1989 governments to address this problem, including a seismic risk classification system, retrofitting programs, warning and preparation campaigns, a public list of structurally unsafe buildings, and an initiative that marked those buildings expected to collapse in the next major earthquake with a large, red, circular sign, colloquially known as a "red dot." Three appendixes elaborate on the 1977 earthquake and disaster response in Romania. These include a discussion of the tangential connection of the earthquake to Ceauşescu's Civic Center Project and its House of the People; a regime-created list of the identified dead; and an earthquake preparation guide.

All the regime's disaster response activities served its objectives of bringing economic activity back to pre-earthquake planned targets and ending what Ceauşescu perceived as the "chaos" and "panic" of disaster response. Many engaged in the disaster response, but Ceauşescu led it, took credit for its successes, and blamed others for its deficiencies. The response activities' initial effect on the regime was, therefore, not to promote solidarity in the face of disaster, but to "increase the gap between Ceauşescu and his colleagues."[3] By 1977 he described himself as the nation's patriarch and his tenure as the "Golden Era." Privately, Ceauşescu angrily reproached most of the Romanian Communist Party (RCP) and government leaders, accusing them of being "utterly confused" after the earthquake struck. As the US Department of State noted, "The earthquake provided Ceauşescu with an unparalleled opportunity to display his complete dominance of party and state activity."[4] From 1965 to 1989, he led one of the most repressive, insular, and nationalist post-World War II communist regimes in central and eastern Europe, with the 1977 earthquake marking its halfway point.

Two motivations guided Nicolae Ceauşescu during the 1977 earthquake disaster response. The first was economic. The influx of foreign assistance in the form of in-kind goods and cash and spontaneous domestic giving served as two separate yet related opportunities to go beyond receiving what was given and proactively make specific asks abroad and at home. Publicly, the regime stated that such efforts were for earthquake relief, but privately they were intended to bolster economic recovery. The regime prioritized economic objectives during the immediate disaster response, too, dispatching firefighters and Securitate officers to address damage to economic assets, search for both survivors and valuables at the collapsed building sites, and retrieve valuables well after search and rescue closed. It actively and clandestinely steered foreign assistance toward cash and specific technical and construction equipment.

Ceauşescu's other motivation—a desire to end the "chaos" and "panic"—emerged as the regime worked through its earthquake disaster recovery. Every day as he traveled through the city he was reminded of the earthquake's disruption: the rubble from the partially and fully collapsed high-rise buildings; the search and rescue teams; the bulldozers moving and dump trucks hauling the debris; the blocked roads and rerouted traffic; people sleeping in their cars and moving their household possessions outside in fear of aftershocks; assessment commissions inspecting the structural damage; and RFE's nonstop programs allowing Romanians to speak live on-air.

Ceauşescu's indecisiveness was apparent throughout the disaster response. Often, he conflicted with underlings and experts about the best course of action. Initially, he favored repair over demolition of the structurally unstable buildings, not necessarily so as to make them safe but rather to avoid the disruption of yet another pile of rubble. During the disaster response, he clashed with scientific experts and regime leaders about a range of things beyond just this issue as well: how to manage the first responders; which engineering techniques were best for search and rescue; whether, during those first days, to remove and inventory household items when people might still be rescued from the rubble; and how to manage foreign humanitarian assistance receipts and deter looting. When experts' informed recommendations conflicted with his ideas, he just disregarded them, rarely engaging their arguments. The regime's disaster response highlights Ceauşescu's mistrust, disdain, and disregard for scientific specialists and credentialed experts.

Elena Ceauşescu (1916–89), Nicolae's wife, shared his mistrust of experts. By March 1977, she held the regime's number two position. Just under them was Prime Minister Manea Mănescu (1916–2009).[5] All government officials in ministry or leadership posts were members of the RCP. Many who led the disaster response were longtime members of the Central Committee of the Romanian Communist Party (CCRCP) and served on the Political Executive Committee (PEC). Two had primary roles in the 1977 response efforts: Bucharest mayor Ion Dincă (1928–2007), in office for nine months when the earthquake occurred, and minister of the interior Theodor Coman (1928–96), in office for two years at the time.[6] Dincă managed the Bucharest-level efforts, especially the teams that assessed the earthquake-damaged buildings' structural safety, while Coman commanded the first responders, which included firefighters, police, and Securitate troops. Throughout the regime's disaster response, Ceauşescu met frequently with PEC members, which included long-serving CCRCP members.[7] While Ceauşescu was ultimately responsible for (and took credit for) the regime's policies and actions, those within the regime drafted, supervised, and carried them out. Engineering experts, architects, and construction workers played especially important roles. They assessed the damaged buildings' safety, made recommendations, drafted plans, and supervised and made repairs. Their earthquake recovery efforts highlight the integrity and pride of professionals and civil servants. Industrial workers bore the brunt of mandated policies that took their labor and donations in the name of the 1977 earthquake victims. Citizens were "encouraged" to help clean up Bucharest, the displaced to find housing on their own, and residents of damaged buildings to make their own repairs. All were ultimately accountable to Nicolae Ceauşescu, who, as president in name and authoritarian in function, was accountable to no one.

Notes

1 INCD pentru Fizica Pământului, "Cutremurul vrâncean din 4 martie 1977."
2 Major earthquakes register between 7.0 and 7.9 on the Richter scale; great earthquakes register 8.0 or higher.
3 Ashley and Munteanu, "Ceauşescu's Leadership Position," 543.
4 Ashley and Munteanu, 545.
5 Mănescu was prime minister, 1975–79; Central Committee of the Romanian Communist Party (CCRCP) member, 1960–79 and 1982–89; and Political Executive Committee (PEC) member, 1974–79 and 1984–89. Secţia Cancelarie, 18/1977, 2, FCCPCR, ANR; Dobre, *Membrii C.C. al P.C.R.*, 138–39.
6 Dincă was Bucharest mayor, 1976–79; CCRCP member, 1972–73 and 1976–77; and PEC member, 1976–89. Coman was interior minister, 1975–78; Communist Youth member from 1947; RCP member from 1953; and CCRCP member, 1969–89. Dobre, *Membrii C.C. al P.C.R.*, 168; 211–212.
7 These members included Ştefan Voitec (1900–1984; CCRCP/RCP member, 1948–84; and PEC member, 1974–84), Aurel Duma (1919–93; CCRCP member, 1969–89; and head of the CCRCP Chancellery, 1972–78), and Nicolae Giosan (1921–90; CCRCP/RCP member, 1960–89; and supplementary PEC member, 1974–89). Dobre, 240, 299, 626.

1 A Primer

Communist Romania under Ceaușescu

At times the communist Ceaușescu regime (1965–89) could be brutal, negligent, ineffectual, or absurd. Life in Romania under the regime shared elements with life in other socialist countries in central and eastern Europe and the Soviet Union but was in many ways unique. This primer covers the most important, unusual, influential, and difficult aspects of the Ceaușescu regime, providing context for its 1977 earthquake disaster response.

Communist Romania's Second, and Last, Leader

Following World War II, Romania entered the communist bloc thanks to decisions made outside of it. In 1939 the Molotov–Ribbentrop Pact (Treaty of Non-Aggression) between the Soviet Union and Nazi Germany created a secret postwar plan that divided central and eastern Europe and gave the Soviets authority over Romania. Before and during World War II, the Romanian Communist Party (RCP) was small, impotent, and illegal. It had barely one thousand members, due in large part to persecutions, arrests, and imprisonments by the Romanian fascist government. As a teenager in the 1930s and a young adult during World War II, Nicolae Ceaușescu was imprisoned alongside others for communist activities, many of whom later became influential communist party and regime leaders. Starting in August 1944, Soviet troops occupied Romania and trained and guided the fledgling RCP, which then, with Soviet help, ousted the fascist Antonescu regime and reinstalled the monarch. King Michael served as a figurehead until 1947, when the Soviets and RCP forced him to abdicate. Following World War II, the Soviets stayed to head the Allied Control Commission, assisting the RCP in forming its government, expanding industrialization, organizing land collectivization, and developing its political police, the Securitate. As the Soviets did so, they introduced procedures and named departments mimicking those of the Soviet political police, the KGB.[1] While the Soviets were instrumental in developing Romania's communist government, they left in 1958. Two Romanian leaders subsequently steered the country: Gheorghe Gheorghiu-Dej (1901–65) led Romania from 1948 until his death, and Nicolae Ceaușescu directly succeeded him, leading the country until his execution in December 1989.

DOI: 10.4324/9781032632933-2

Important to understanding how Ceauşescu led his regime is an awareness of how his predecessor and mentor showed a liberal face outside Romania but instituted terror and repression at home. Ceauşescu is often given credit for Romania's unique economic and ideological separation from the communist bloc, but Gheorghiu-Dej inserted the first wedge between Romania and the Soviet Union. In 1958, two years after Soviet chairman Nikita Khrushchev (1894–1971) made his 1956 "secret speech," the first official criticism of Joseph Stalin's (1879–1953) political terror, Gheorghiu-Dej oversaw the withdrawal of Soviet troops from Romania. He focused on internal economic policies, reverted the Slavicized spelling of the country's name to its Latin form, released prominent historians from prison, and broke from the Soviet bloc when voting on certain inconsequential issues at the United Nations.

Gheorghiu-Dej's regime used targeted terror. He imprisoned intellectuals, clergy, and political opponents in Romania's own gulag archipelago, which comprised almost three dozen secret and not-so-secret prisons, or sent them to dozens of industrialization projects that used prisoner labor. Across the communist bloc, hard labor was used in large-scale development and modernization projects not only as punishment but as part of a larger socialist undertaking to transform people through labor. Marxist theory holds that a key component of human evolution is the ability to labor using tools and that through labor humans can, and will, become the best version of themselves. Regardless of whether Gheorghiu-Dej ascribed to this theory, in 1949, just after he took office, his regime began sending prisoners to work on the massive Danube–Black Sea Canal project. By the early 1950s, as many as sixty thousand forced laborers had been sent to work on this project, thousands to their deaths.[2] Overall, it is estimated that during Gheorghiu-Dej's regime, half a million people died due to imprisonment, overwork, starvation, or exposure to the elements.[3]

Gheorghiu-Dej met Ceauşescu during World War II when they both served time in Târgu Jiu Prison, imprisoned alongside hundreds of others for their RCP activities and membership. Gheorghiu-Dej took Ceauşescu and other young prisoners (many of whom would later rise in the RCP ranks) under his wing. He recognized something in Ceauşescu's potential and commitment to communism and appointed him to important leadership positions, shepherding his rapid rise within the party. "For Gheorghiu-Dej, Nicolae Ceauşescu was the perfect embodiment of the Stalinist apparatchik. He appeared to Gheorghiu-Dej as a modest, dedicated, self-effacing, hard-working, and profoundly loyal lieutenant."[4] Ceauşescu held high office in the Ministry of Agriculture during collectivization and then served as second in command in the Ministry of Defense. And so, when Gheorghiu-Dej died suddenly from cancer in 1965, after some negotiation with other top leaders, Ceauşescu succeeded him. By the time of the 1977 earthquake, Ceauşescu had given himself an array of positions with grand titles: president of the Socialist Republic of Romania, general secretary of the RCP, president of the Council of State, chairman of the Socialist Unity Front, chairman of the Defense Council and supreme commander of the Armed Forces, chairman of the Supreme Council of Economic and Social Development, and chairman of the Ideological Commission of the RCP. Over the course of his twenty-four-year reign, Ceaușescu, at first popular, became notorious for his cult of personality, paternalistic state, repressive hold on power, strange behavior, nationalism, and lavish lifestyle, built his power base, shaped a neo-Stalinist regime, placed his family and loyal followers in influential positions, exploited nonsocialist countries' appreciation, used fear to stem domestic opposition, and drastically changed day-to-day life in Romania through the ever-present political police and harmful social and economic policies. But Ceaușescu did not begin his rule as a despot. At first, he tried to shape Romania's government, economy, and culture to reflect his socialist vision, a modern, industrialized one with a particular ethnic Romanian nationalist form.

By the 1977 earthquake, Ceaușescu had fully consolidated his power and established his authoritarian rule in a regime since characterized as "dynastic socialist" because of his family's powerful roles. Elena Ceaușescu, his wife, became the most powerful among them. Of their three children, eldest son Valentin (b. 1948) and daughter Zoia (1949–2006) stayed out of government and the regime, while youngest son Nicu (1951–96) held several positions in the RCP and government in the 1980s but was preoccupied with the distractions available to him because of his privileged position. In a paternalistic dynastic fashion, Ceaușescu had plans to pass his rule onto Nicu, yet it was Elena who held significant power in the regime (see Figure 1.1).

Figure 1.1 Nicolae (left) and Elena Ceaușescu on his sixty-second birthday, January 16, 1980.

Source: Fototeca online a comunismului românesc, K006, Cota NA.

By 1977 until her execution in 1989, Elena Ceauşescu was a full member of the regime's Political Executive Committee (PEC) and essentially second in command. Like her husband, she had a storybook communist pedigree: born of rural, simple, ethnic Romanian and Orthodox Christian stock, she left school at fourteen and worked in a textile factory. She promoted an image of herself as the country's matriarch and a credentialed scientist. But, as most now understand, her PhD in chemistry and published research papers were written either completely by or with the help of ghostwriters. Her claim to hold honorary membership in the Illinois Academy of Sciences was not true.[5] She used her position in the regime to serve herself and buffer her husband from those around them whom they both often mistrusted. Together they created a unique neo-Stalinist, dynastic regime, with the patriarch at its center and his spouse unequal to him but powerful in her own right.

"Maverick" to the West, Despot at Home

Early in his tenure, Nicolae Ceauşescu showed a Janus face, touting reforms for an audience abroad while implementing repressive policies at home. By the 1970s, the nonsocialist world saw Ceauşescu as an open and sympathetic figure, a potential wedge in the communist bloc. Ceauşescu became the West's "Maverick of the East" after he broke from the Warsaw Pact in 1968, when he publicly criticized the Soviet invasion of Czechoslovakia, calling it "a great mistake and a grave danger to peace."[6] He presented himself to the West as a potential ally inside the communist bloc, open to Western approaches, while still exalting his commitment to socialism. As a direct result, there were more frequent diplomatic, economic, and cultural contacts between Romania and the West. In 1969 US President Richard Nixon visited Romania, and in 1973 Ceauşescu reciprocated with a visit to the United States. Romania joined several important economic organizations, including the General Agreement on Tariffs and Trade (1971), the International Monetary Fund and the World Bank (1972), and the European Common Market (1973), and the United States granted it most-favored-nation status (1975). Nonsocialist entities extended credit and offered low-interest loans to support Romania's economic transition from consumer production to heavy industry.[7] By the 1970s, Ceauşescu was touting separation from Soviet—and Russian—spheres of influence, erasing inconvenient historical precedents, exploiting "Romanian" nationalist sentiment, and supporting projects to demonstrate Romanian independence from and superiority over others in the communist bloc. The regime stepped up natural resource extraction and exports in exchange for technological expertise and started new joint ventures with Western companies, a break from the exclusive economic joint ventures with other socialist states. But while Ceauşescu was seen by the West as a bright hope, at home there was a very different view. As the CIA reported in August 1977, "Ceauşescu remains Eastern Europe's most unusual Communist leader, integrating nationalism and Marxism in a uniquely Romanian blend. His domestic policies in recent years … increasingly have departed in significant ways from prevailing Soviet and East European norms."[8]

Through the late 1960s and early 1970s, Ceauşescu allowed Romanians some nominal freedoms. Consumerism and nationalism increased and he lessened travel restrictions and terror tactics. Starting in the penultimate decade of his rule, however, he clamped down on some of the most important aspects of day-to-day life while promoting a cult of personality for himself, Elena, son Nicu, and his extended family. The West looked the other way as his regime's repressive domestic policies and practices increased and intensified over the 1970s. These policies scripted cultural expression, limited and controlled access to news and entertainment, restricted international travel, controlled employment through the ever-present threat of dismissal, and criminalized homosexuality, abortion, and contraception. These restrictions were not always fully enforced and Romanians cleverly worked around and through them, but they still contributed to a domestic environment unique in the communist bloc.

Just a year into his reign, in late 1966, Ceauşescu criminalized contraception and abortion in almost all circumstances, a brutal, cruel, and dystopian policy that unsuccessfully aimed to raise the birth rate.[9] Abortion had been legalized in 1957, which had led to a reduction in the birth rate and relaxed earlier restrictions that allowed abortion only when a pregnant person's life was at risk. Reversing course, the regime promoted its anti-abortion law as "pronatalist." It held criminally liable pregnant people who sought out, attempted, or successfully had an abortion and those who assisted them. Officials performed invasive physical checks on menstruating people, often at their workplaces, to "catch" whether someone was pregnant. One might find contraception on the secondary market or receive it through care packages from abroad, but officially all forms were considered illegal contraband. To promote the policy's objective of more live births, the regime claimed reproduction as a productive output of its laborers in the socialist economy and declared those who birthed and raised many children to be "heroines of socialist labor." Awards abounded: the Maternity Medal was given to women with five or more children, and Orders of Maternal Glory were awarded for seven, eight, or nine children (assigned to the third, second, and first order, respectively). The highest distinction, Heroine Mother, was reserved for those who birthed *and* raised ten or more children. These awards came with some benefits, including a one-time payment, invitations to special events, days off work, and a certificate. Only during the first year of the anti-abortion law did birth rates rise. The remaining years saw fewer pregnant people and significantly more deaths from abortion attempts.

State visits to China and North Korea in 1971 inspired Ceauşescu to make significant changes in his rule. Awestruck by the displays of power each authoritarian brand demonstrated, he mandated upon his return that cultural outputs align with rigid aesthetic prescriptions, increased overt demonstrations and activities that lauded him and Elena, and gave her a more active and influential role. He announced these changes in a speech that came to be known as the July Thesis. Ushering in "his version of a Maoist cultural revolution," Ceaușescu set policy priorities that moved against cultural autonomy, imposed guidelines requiring aesthetic decisions to follow socialist realism and Romanian nationalist forms, and punished cultural producers who did not abide.[10] He outlined seventeen mandates that gave more power to the RCP, intensified political-ideological education in schools, instructed the state-run media to promote his ideology, and resurrected certain banned books and authors. The July Thesis refined Ceaușescu's vision, one that returned to industrialization, promoted Romania's "social and ethnic homogenization," aimed to maximize domestic resource use, promoted himself as the primary symbol of party and national unity, stressed "neutrality within the world communist movement," and reestablished "cordial relations" with communist political parties in some nonsocialist countries.[11]

Also arising after this trip to China and North Korea was the Ceaușescu cult of personality, through which Ceaușescu became "not an authentic leader or an example to be emulated but an idol to be obeyed."[12] He saturated Romania with the Ceaușescu brand, demanding increased participation in parades and spectacles, hanging banners about his and the regime's achievements, displaying the ruling couple's portraits in all state-run buildings, creating a new state holiday to observe his birthday, and flying the flags of the Socialist Republic of Romania and the RCP in public buildings, including schools, factories, and cultural centers. Buildings constructed during his tenure were marked with plaques that read "Of the Nicolae Ceaușescu Era," and places visited by the ruling couple were required to display a plaque that read:

> The great honor and privilege of receiving as guests Dear Comrade Nicolae Ceaușescu, Secretary General of the Romanian Communist Party, and Comrade Elena Ceaușescu, eminent leaders, who dedicate their heroic revolutionary actions to the well-being and happiness of the Romanian people.[13]

Ceaușescu created and encouraged laudatory descriptors of himself, Elena, and the regime. "His tenure as president was known as the 'golden era of Ceaușescu.' The state-run media embellished all references to him with such formulaic appellations as 'guarantor of the nation's progress and independence' and 'visionary architect of the nation's future.'"[14] Elena Ceaușescu, too, changed after their 1971 Asian states visit, in particular, due to the political power she saw held by Jiang Qing, Mao Zedong's fourth wife. She witnessed it in greater force during a December 1976 trip to China, just after Qing and the others in the Gang of Four contested Mao's successor. Nicolae had already deputized Elena into the regime's power circle months before, and, on her return in January 1977, he formalized that recognition.[15] Her power play was visible to all at the time:

> Initially described as a mere shadow of her husband, the Romanian First Lady now plays a significant role within the Romanian decision-making process.... She has managed to build a parallel cult of personality to her husband's, and professional party minstrels compete to compose paeans to the mother of the nation.[16]

Propaganda played an important role in enforcing these messages. In what would be characterized today as "fake news," Romanian domestic, state-run media first and foremost served the regime's ideological mission. The Ceauşescu regime was not unique among the communist bloc in its use of self-serving state media. It controlled all domestic media: national and local daily and weekly newspapers and magazines, two television stations, three radio stations, the journalists who worked for them, and the physical infrastructure where they were produced. By the end of the Ceauşescu era, television programming was limited to two hours per day, showing only domestically produced programs, RCP-approved films, and footage of the ruling couple's official activities. It ran banal yet practical information, too, like TV viewing guides, weather reports, and classified advertisements. Ultimately, it aimed to mobilize all "news" to serve the regime's objectives and promote its ideology.

Romanians did not passively live with media control but actively filled that entertainment and information vacuum. Even though foreign programming was illegal, by the mid-1980s, more than half of Romanians listened to foreign radio programs via shortwave. They had the most VCRs per capita in the communist bloc and passed around shared tapes of Western movies. In 1980 the Ceauşescu regime increased its surveillance of foreign radio, in particular Radio Free Europe (RFE). Political police working in the operation code-named "The Ether" (Eterul) traced who sent letters, special reports, and other correspondence to RFE, whether the station broadcast them or not, and surveilled those who had any contact with or were related to RFE Munich staff.[17] RFE frustrated the regime because Romanians loved and trusted it. One thank-you station broadcast illustrates the importance of RFE: "Long live the traitors at Radio Free Europe for the splendid service they are providing the Romanian nation."[18] To track such correspondence to RFE and others, in 1983 the regime outlawed owning any kind of duplicating equipment except one typewriter, which, if owned, had to be registered with the police (who collected a typed sample from it). This typewriter ban, officially the Duplication Control Decree, mandated that the Securitate "keep track and execute control over the production and use of record keeping and duplicating equipment and typewriters" and allowed only "socialist units"—that is, government departments and agencies—to have copy machines, typewriters, and other duplicating or printing devices.[19]

Foreign-produced media outlets were among several "outsiders" that the regime followed. Ceauşescu feared criticism, opposition, dissent, or threats. The regime defined an outsider as someone who was not ethnically Romanian, not Orthodox Christian, or not committed to the RCP's objectives. This privileging of all things ethnic Romanian denigrated ethnic and religious minorities, casting them as outsiders within the confines of the ethnic Romanian socialist state. While the Romanian population was majority ethnic Romanian and Orthodox Christian, a diverse range of ethnic and religious groups lived side by side. The same is true today. The largest minority groups, who together represent almost 10 percent of the population, are ethnic Hungarians and Roma. During the Ceaușescu era, the regime paid particular attention to Roma, to the point of demonization, referring to them in one example as a "parasitic criminal element" alongside panhandlers, vagabonds, and prostitutes.[20]

The regimes under both Gheorghiu-Dej and Ceaușescu charged the Securitate with keeping tabs on such "outsiders," concentrating its surveillance efforts on non-Romanian ethnicities, non-Orthodox religious followers (especially the small Jewish community), foreigners, Romanians who had contact with foreigners, and former members of Romania's World War II-era fascist regime and those who had collaborated with them. The regime even considered orphans and other "unattached" people as potentially dangerous and worthy of surveillance. As for foreigners, the political, economic, and cultural openness allowed the entrance of scholars and students from the socialist and non-socialist world, whom the regime also tracked. This obsession with outsiders was an expression of Ceaușescu's form of Romanian nationalism. He saw ethnic, religious, and regional differences as subordinate to those of the nation while at the same time defining Romanian nationalism as ethnically Romanian and Orthodox Christian, to the detriment of all "outsiders."[21] Communist ideology fueled his goals. He was committed to "communist idealism" and was obsessed with Romanian modernization in all its forms, expressed most assertively in the state's control of the economy and Ceaușescu's interest in the "built environment of socialism."[22]

Urbanization Efforts

In 1972, to support its modernization goals, the Ceaușescu regime instituted systematization (*sistemizatizare*), or "any kind of physical or territorial planning, and the actions resulting from this planning."[23] This territorial restructuring moved rural residents to urban areas, located factories in towns near populated areas, and transformed village clusters, or communes, into "agro-towns." It aimed to raise living standards and increase population density in rural areas by building multi-unit housing, improving transportation, building new factories and staffing them with peasant labor, and setting up social services.[24] Romania's urbanization under systematization differed from the "new town" plans of other communist countries in that it aimed to transform rural areas and not necessarily build new urban ones.

The territorial restructuring went hand in hand with shifts in the economy. After World War II, Romania's economy shifted from one that was overwhelmingly agricultural to one almost equally shared between agriculture and industry. Previously, most Romanians had worked in agriculture, tilling small tracts of land in the plains, shepherding sheep and goats in the mountains, cutting timber in the forests, and fishing in the Danube Delta and the Black Sea. During the first decade after World War II, the Romanian communists collectivized agricultural land, built factories, expanded resource extraction, moved people to the cities, and constructed hundreds of thousands of apartments for the new urban workforce.

Across Europe, including in the communist bloc, a severe urban housing shortage followed the war. With increased industrialization, people migrated to cities. The rapid increase in population density in urban areas, from systematization and the postwar baby boom, contributed to the significant housing shortages.[25] Before systematization, many Romanians who worked in urban factories had commuted to the cities from their rural homes. This changed, however, as the communists modeled new housing projects on the Soviet and British models of clustered apartment "blocs" or housing estates. Ceauşescu increased the annual new apartment construction targets significantly from those of his predecessor. The Gheorghiu-Dej regime constructed fourteen thousand new apartments annually in the 1950s and forty-five thousand in the following decade. This paled in comparison to the 1970s under Ceauşescu, with the construction of as many as one hundred thousand new apartments annually.[26] Under Ceauşescu, marriage would get citizens on a list for an apartment, which might allow them to leave dormitory housing or receive a residency permit, as most people needed state approval for where they lived.[27] More likely than not, however, young couples shared an apartment with one (or both!) of their families. In addition to telling citizens where they could live, the regime prescribed how much space they could have, approving 8 square meters per person, equivalent to a 32-square-meter apartment shared by four.

The regime eventually expanded its systematization efforts to include cities. In Bucharest, this included constructing new apartments, expanding the number of administrative districts from six to eight, building larger thoroughfares and boulevards, digging the city's subway tunnels, and incorporating surrounding villages into the city. The 1977 earthquake disaster response justified the regime's amplification and fast-tracking of these efforts and the resurrection of an interwar-era plan to build the new Civic Center Project (Appendix A). Within weeks of the earthquake, Ceauşescu announced his vision and plan for its construction, adding its centerpiece, the House of the People, a political-administrative center with a grand assembly building. In 1981 the regime laid its cornerstone, and the decade saw massive numbers of building relocations, resident evacuations, neighborhood demolitions, and domestic raw materials and labor diverted for its construction. These urban and rural demolitions in the name of systematization continued well into the 1980s, drawing international attention and scorn. As the historian Dennis Deletant explains, the systematization plans were "entirely directed by the [communist] party," and those professionals (architects and engineers), as well as the residents affected, were completely left out of its development, with the grim exception of evacuating or demolishing their own homes.[28] Ceauşescu saw industrialization and modernization as the means to achieve communism and sought to transform urban, rural, industrial, residential, and governmental spaces to reflect that ideal. Yet, for Ceauşescu, the socialist ideal was national, not international, Romanian, not universal. He considered an individual's interests and freedoms subordinate to those of the group. He was primarily concerned with forging the "socialist man" into a Romanian nationalist form.

Professionalization and Expansion of the Securitate

Once in power, Ceauşescu reformed the Securitate, one of the regime's most important institutions. It did not work as "a monolith with a single overriding intention—the opinion of most Romanian citizens—but as a multicentric organization fragmented among many parts."[29] It was not a "secret" police force; rather, its actions and presence were relatively out in the open, saturating daily life through the activities of its officers and willing and unwilling informants.

The Securitate's objective was to gather, distribute, use, and attribute value to domestic and foreign information about all manner of things, which it assembled into files, the first great socialist industry.[30] These files contained information Securitate staff gathered themselves or from informants. File creation was akin to feathering a nest for no particular bird. Officers and informants gathered and fabricated information to fill the pages that populated the files. The extant Securitate files are not trim edited documents; they are full of redundancies. Officers and staff typed up handwritten notes and then retyped them again. They transcribed recorded or live telephone conversations. They photocopied those transcripts. If the transcripts were handwritten, they might type them up and then photocopy those documents, too. They archived the files. Periodically, they purged a selection, or sometimes all, of them. The complication, redundancy, and "busy work" were intentional. The rationale of the duplication and splintering was to limit staff from knowing who was working on what, a practice known in Romanian as *conspirativitate* and commonly used by political police.[31] Restricting officers and others from having a clear understanding of how information was gathered and from whom was a means of protecting the organization and its staff. This redundancy was a hallmark of the post-Stalinist communist bloc's security forces and agencies. In Romania, files were intended to be read only by the Securitate, Nicolae and Elena Ceauşescu, and PEC members.[32]

All that information was gathered by different regional departments (or directorates), each with its own units and special units with distinctive objectives and responsibilities. "Unit T" listened to conversations in "bugged" locations; "Special Surveillance Unit F" followed people; "Service S" read and censored mail. Directorates were divided by specialty: I was concerned with internal information, II on economic espionage, III on counterespionage, IV on military counterespionage, and V on the protection of regime leadership.

New units and special units were created to perform specific tasks as the need—or whim—arose.[33]

One structural change brought the Securitate under a newly created Interior Ministry. In 1972 the Securitate's precursor, the Council for State Security (formed under the Soviet Allied Control Commission), merged with the Ministry of Internal Affairs to form the Interior Ministry.[34] In 1973 the Interior Ministry became the umbrella organization for the Securitate, firefighters, police, and penitentiary workers. Each department had its own education and training units, hospitals, cultural institutions, cinemas, publications and printing houses, a special center for "psychophysiological" research, laboratories, and shops for equipment repair.[35] The Securitate's 1970s reorganization also focused on its personnel as a way to restore its legitimacy. As part of the reforms, Ceauşescu fired almost a third of the Securitate's officers and hired ethnic minorities in equivalent proportions to their representation in the population, as well as recruiting applicants with higher education. By 1989, a mid-level Securitate officer earned about $400 a month: double the average Romanian worker's salary, and 13 percent more than the equivalent ranking army officer.[36] In addition to higher pay, Securitate officers had access to special shops, which sold goods unavailable to most, and to special facilities, such as exclusive restaurants, bars, and spas. Women worked in the Securitate, too. However, because of the norms of the time, they could not do the same work as men (like sitting alone in a restaurant or drinking with an informant) without garnering unwanted attention, relegating most women to work as secretaries and typists.[37] The Securitate also issued a professional journal for officers and supported the publication of pulp fiction spy novels with sympathetic Securitate protagonists.[38] By the last decade of the Ceauşescu regime, Romania had the highest number of political police officers per capita of any communist country in Europe (one for every thirty people), employing thirty-nine thousand officers for a population of twenty-three million.[39]

Ceauşescu reformed the Securitate to consolidate his power and present a more professional, less brutal organization, one under his control. As Deletant explains, "Ceauşescu was clever enough to realize that his own position of personal dominance was ultimately dependent on the loyalty of the Securitate."[40] He also used the public's knowledge of the Gheorghiu-Dej regime's use of terror to gain credibility. Within months of being named Gheorghiu-Dej's successor, Ceauşescu denounced the illegal imprisonments and forced labor under his predecessor, claiming that the Securitate was solely responsible for the forced labor, imprisonment, and targeting of ethnic and religious minorities during Gheorghiu-Dej's tenure. He claimed, too, that the Securitate had acted on its own when it tortured people to extract false confessions, which were then used in show trials that led to defendants' imprisonment or execution.[41] In 1968, Ceauşescu rehabilitated Lucreţiu Pătrăşcănu, a communist leader who was executed in 1954 after a show trial. This belated justice helped promote Ceauşescu as the leader who could rein in the Securitate and curb its practices. A practical result was that the Securitate no longer needed to actively engage in terrorizing activities because knowledge of its ability to do so was a powerful repressive measure in and of itself.[42]

In their understanding of the Securitate's power, Romanians moved from experiencing terror to fearing it. Even after the 1970s Securitate reforms, Romanians feared being arrested and thrown into prison, "despite the fact that the regime had already changed its methods to control from randomly applied repression to extensively disseminated fear."[43] The Ceaușescu regime encouraged this atmosphere of fear by having its Securitate officers and informants work relatively out in the open. One could expect their presence at work or school, at church, among one's neighbors, or even among friends and family. The Securitate paid special attention to "outsider" groups and those who had possible or real contact with foreigners—students, scholars, business people, and workers. It did so inside and outside Romania. It is widely understood that in the 1980s Securitate officers or informants orchestrated assassinations like that of dissident Gheorghe Ursu, used radiation to hurt RFE staff, supported Venezuelan Carlos the Jackal's bombing of the RFE Munich office, and cracked down on the worker protests in Brașov and the revolutionary December 1989 protests across Romania.[44] Under Ceaușescu, the Securitate's objectives morphed from "destroy the class enemy and prevent infractions against state security" to "defend national values." Ceaușescu declared his "new mission for the Securitate: 'war of the entire people.'"[45]

Still, the Securitate's primary objective was to use surveillance to gather information for the regime. Throughout the communist era, the severity with which the Securitate applied its methods changed, while its primary function remained the same: "to collect valuable information directly from the source, about citizens' attitudes about the regime in order to prevent and repress any actions not in line with the official ideology."[46] To do so, it supplemented the work of officers with information from willing and unwilling informants.

Anyone could share information with the Securitate, but it primarily relied on an informant network that grew exponentially through the end of the Ceaușescu regime. Before he came to power, the estimated number of informants (active and inactive) was 80,000; by 1967 it was 110,000, and by 1989 it was at almost half a million.[47] These individuals were organized into four categories: "informants" (*informatori*) met with Securitate officers and gave information; "supporting people" (*persoane de sprijin*) spied on specific people, things, or events; "residents" (*rezidenți*) established contacts with people to collect information; and "meeting place hosts" (*gazde ale caselor de întâlniri*) allowed the Securitate to use their home, workplace, or other location for meetings, but did not take part. To recruit informants, an officer might gather information about them to entice or threaten them into cooperation. Valuable information about their employment could include whether they had been sanctioned or had asked for a transfer. Their personal information might highlight whether they had requested a new or larger apartment, applied for a passport, or needed medical treatment not readily available.[48]

But first, an officer would create a potential informant's "story" (*legenda*) from professional and personal information, potentially including a general biography; information the Securitate already held, including any denunciations; recorded telephone conversations; photographs, especially with anyone known to be "hostile" to the regime or a potential lover; or evidence of contraband or items considered hostile to the regime.[49] Once the story was assembled, an officer might approach a potential recruit with incentives like a larger apartment or one in a better neighborhood, reduced waiting time if in the queue to buy a car, better placement for children in university, or advancement or perks at their job. Officers also recruited political detainees and prisoners (who were usually followed after release even if they did not become informants). Securitate officers would often approach prisoners during their final days of detention, with the incentive of a return to "normal" life or the threat of longer detention if they did not cooperate. The Securitate also lauded informing as a patriotic act.[50] The first meeting with a potential informant would be presented much like a celebration or party, with the officer well dressed, offering celebratory food and drink. If the prospective informant agreed, they received a code name and signed a loyalty oath.[51]

What followed was not predictable. Meetings were not mandatory or regular.[52] There was no salary; rather, informants might receive gifts like cash, foreign goods not readily available, or favors. Informants inside prison might "earn" packages and letters from family, food, clothing, or medicine. Most commonly, informants or their family members received professional help. This could be a job transfer or promotion, assistance with academic or artistic publications or productions, or inclusion in a foreign delegation. Personal help might include a passport, permission to travel, or home repairs.

Indicative of the complicated relationship everyday people had with the Securitate, informants were both willing and unwilling, and they informed both rarely and frequently. They were used by the regime and used it. They were believed and doubted. They were coerced to say things. There was no one type of informant or single reason for informing. Smaranda Vultur describes Securitate informants:

> Some were gentle and sympathetic, others were zealous and intent on harming their targets and exaggerating their deeds, creating an image of them that corresponded to the Securitate's expectations. Some informers were spiteful and envious, others were stupid or naïve, while still others almost transformed this practice into a perverse hobby. Some informers were talented writers who structured their stories, building up suspense which culminated in incriminating arguments like a sort of [a] punch line, even though they were based on presuppositions.[53]

Informants were released if they had poor results, skipped meetings, provided unclear, misleading, or wrong information, refused to collaborate, or outright lied. Legally, one could refuse to collaborate with the Securitate. People said no, or they didn't say yes, with few consequences. So why, by the end of the Ceauşescu regime, did so many people inform? The benefits and repercussions in one's personal and professional life could be substantial, and the repercussions are real. But they were difficult to predict.

In addition to domestic and foreign information gathering and file creation, the Securitate had a primary role in satiating the regime's "hunger for hard currency."[54] The Gheorghiu-Dej and Ceauşescu regimes prohibited citizens from holding or using foreign currency, yet both actively did so for external trade, tourism, and clandestine operations. The Securitate's Directorate for Foreign Intelligence (Direcţia de Informaţii Externe, DIE) supervised the Special Foreign Currency Operations Department (Compartimentul de Operaţiunile Valutare Special, OVS), reorganizing it as the Special Foreign Currency Shares (Acţiuni Valutare Speciale, AVS), which managed the foreign currency deposits, payments, and transfers. During the 1970 and 1975 flood and 1977 earthquake disaster responses, the Ceauşescu regime looked to foreign sources of cash and charged its Securitate with sourcing them.[55] The Securitate's motivation went beyond national duty to blatant self-interest: starting in the mid-1950s, the Securitate kept for its operations 20 percent of the foreign hard currencies it obtained, earned, or acquired through its business or other transactions.

The Securitate under Ceauşescu oversaw and even ran for-profit businesses to earn foreign hard currency. Beginning in the 1970s, it was involved in several Romanian-foreign joint ventures, where the regime took "confidential commissions" in exchange for granting foreign firms the right to do business in Romania, share earnings, or launder money through transitional accounts. For example, in the 1980s the Swedish furniture maker IKEA agreed to pay these "commissions" (i.e., kickbacks) on top of the regular wholesale price it paid for furniture produced in Romania.[56] It appears that these business dealings, while clandestine, did not involve illicit activities like prostitution or narcotics trafficking. While the regime used foreign currency, it prohibited most Romanian citizens from doing so, in part to control its use in the informal, secondary market, and other illicit activities. However, Romanians had another valuable currency: foreign cigarettes. They used cartons, packs, or even single cigarettes to buy items, with the US brand Kent being the most valuable.[57] In response, the regime managed this "currency" by monopolizing cigarette imports, with cartons distributed to specialty shops for regime elites.

The Securitate lived, in a sense, in its own world, one that mirrored the regime's objectives. Who was the current enemy? The Securitate would report on them. What group was the most active at this time? The Securitate would gather information on them. Yes, group X is discussing Y among themselves and with others. Daily life became theater, with threats created, enemies fabricated, and risks amplified all to keep the information flowing for even more file creation. Never knowing exactly what these files contained, regime members recognized their potential harm. During the December 1989 revolutionary events, some with access destroyed documents held in the RCP, government, and Securitate archives. Much, but not all, was saved, however—in some cases dug up from holes or fished out of the river.[58] Today, people can request their Securitate file and accredited researchers can access those approved for research at the National Council for the Study of the Securitate Archive (Consiliul Naţional pentru Studierea Arhivelor Securităţii, CNSAS), which holds at least twenty-six kilometers of documents.

Multiple Crises Faced in the 1970s

The 1970s was a decade of natural, political, and domestic crises in Romania. The Ceauşescu regime dealt with the aftermath of two extreme floods and a major earthquake, the exposure of its human rights abuses, the defection of a top regime official, and a strike by more than 30,000 miners. First in 1970 and then again in 1975, the Ceauşescu regime engaged in significant disaster response following severe spring floods. The May 1970 flood waters affected more than 1,600 rural communities, submerging a total of 460 villages or towns, destroying an estimated 40,000 homes, and damaging 11 percent of agricultural land. Internally, it was the first test of the regime's disaster response capabilities; externally, it was the first post-World War II disaster inside the communist bloc to garner attention from socialist and nonsocialist countries alike.[59] In contrast, across the communist bloc and particularly in the USSR, information about disasters like floods and earthquakes was kept within the Iron Curtain.[60] Romania's openness to the West facilitated the outside help it received after disasters, and its significant diaspora, especially in West Germany, sent private humanitarian aid, with new postal agreements making the process of sending packages easier.[61]

In 1977, in addition to the earthquake, the Ceauşescu regime also dealt with two other significant concerns. The first was in February, when dissident Paul Goma wrote about the Ceauşescu regime's human rights abuses in a letter supporting Charter 77, a document drafted by Czechoslovak intellectuals, writers, and other activists that outlined the Czechoslovak government's violations of human rights treaties it had signed. Goma sent his letter to a Charter 77 founder, who gave it to RFE for broadcast.[62] In response, the Ceauşescu regime placed Goma under house arrest, allegedly for violating the law that outlawed criticism of the Romanian state in foreign media, and then in November allowed or forced him to emigrate to Paris. Following the "Goma affair," in July 1978, the regime was shaken when Ion Mihai Pacepa, who served as Ceauşescu's personal advisor and deputy director of the DIE, defected, becoming the communist bloc's highest-ranking official to do so. In response, Ceauşescu fired, demoted, removed, or transferred scores of regime officials and restructured the Securitate once again, out of fear that many of its members may have been involved with or helped Pacepa.[63]

The second significant concern of 1977 was a large protest in August. Before Romania's 1989 revolution, Romanians rarely organized large-scale protests against the regime. Coping mechanisms were private, relegated to personal conversations and the exchange of jokes. But in August, almost ten thousand miners in Transylvania's Jiu Valley went on strike, protesting a reduction in their pension and disability benefits. Days into the strike (the first large protest against the Ceauşescu regime), government officials met with mine representatives and local officials, but absent Ceauşescu, upsetting the miners even more. A request was made, and Ceauşescu arrived from Bucharest, speaking for hours to thousands of miners gathered outside. They began to boo and heckle him. Ceauşescu got flustered, and, most likely because of the large group of miners and his small security detail, he conceded to some of their demands, but ultimately broke most of what he promised. The state-run media never reported on the strike. The Securitate interrogated strike leaders and miners, then moved about four thousand of them, some to the Danube–Black Sea Canal labor camps.[64] Even without official reports of the strike, of course, Romanians learned about it. Public knowledge of the strike and the regime's silent response deterred future protests and isolated strikers, who were frustrated by their limited success. This deterrence-through-denial approach was common: "if no opposition to the regime was reported, then most of the public not only assumed that there was none but, guided by this assumption, questioned the point in displaying any."[65]

Crises Created in the 1980s

The final decade of Ceauşescu's rule and communism in Romania was unique, harsh, and, for many, horrifying. Those outside of Romania remember the communist era under Ceauşescu mostly for the events of its last decade, not primarily because they were the last but because they were a violent outlier to what was happening across the communist bloc. From state-imposed austerity to orphaned children dying of AIDS, and culminating with the only bloody revolution in eastern Europe, 1980s Romania is remembered for the enormity of these events.

Romania was not immune to the global economic crises of the late 1970s and 1980s. Iran raised the price of crude oil, including to Romania, and the Soviet Union stopped its oil subsidies for the bloc. Even though an oil producer since the inter-war era, Romanian heavy industry relied on the crude it extracted and imported. Romania's Western debt came due—$10.2 billion then, about $34.5 billion today—and so Ceauşescu, after trying to restructure it, decided to pay it off. To do so, the regime imposed a severe austerity policy that required almost anything of value produced in Romania to be sold for export, leaving very little for domestic consumption in terms of food, fuel, or consumer goods.

Under austerity, Romanians lived with the daily challenge of finding necessities like eggs, gasoline, and toilet paper, let alone "luxury items" like fruit, chocolate, or coffee. Energy was rationed, too, with electricity and hot water often cut during the day. Salaries were reduced—by the mid-1980s, some as much as 40 percent—but even if one had money, there was very little to buy.[66]

Humor was one way people coped, with many jokes featuring the popular character Bodo. One made light of austerity:

> Bodo was in school one day and his teacher says, "Bodo, come up to the board and draw a pig." So Bodo comes up to the board very confidently. He outlines the snout, two ears, four feet, and a tail and goes back to his seat. His teacher was incredulous. "I said, come up and draw a pig and you only gave me these appendages." Bodo replied, "You didn't say you wanted an export pig."[67]

Austerity was, as Peter Siani-Davies explains, "a massive campaign of export promotion and import substitution … in which the most marketable products, especially foodstuffs, were shipped abroad, leaving only 'patriots' such as pig's feet at home."[68]

Austerity required new ways of doing things. Procuring practically anything meant spending lots of time in queues and "gradually expropriated Romanians of much of their control over time."[69] Lining up for food, fuel, and other consumables, completing household tasks in the middle of the night (when electricity and natural gas were available), and reducing socialization with family and friends because of the inability to procure food and drink were all ways in which austerity policies took not only goods for export but also people's time and energy at home. Working-age adults had little time for supplemental income activities, like tutoring for extra cash. Instead, people scavenged, increasing the power of workers who had access to hard-to-find items. Enterprising grocery workers hid goods behind the counter to sell to or barter with people who could pay them extra (in addition to paying the shop!). Store clerks and restaurant waiters pilfered what they were able to sell on the secondary market or use themselves. Urban dwellers received produce and meat from rural relatives, who had skimmed it from their state-required production quota. It is estimated that the informal, secondary market in the 1980s accounted for more than one-third of Romanians' income.[70] These kinds of scavenging activities the regime considered criminal and in violation of the communal objective of sacrifice for the greater good. It expended a lot of time and energy to surveil the population, deter such activities, and penalize those who engaged in them. In October 1988 Katherine Verdery observed this scavenging in the village where she was conducting anthropological fieldwork. Her description of the activity offers a rich example of the extent to which many worked to procure what they needed for themselves and their families to survive:

> Both early in the day when I arrived and late at night when I left, local authorities were out in the fields with those workers they had managed to round up for the potato and corn harvests, and the streets were crawling with policemen shining powerful flashlights on every vehicle that might divert corn or potatoes into some storehouse other than that of the collective farm. Whether on that night or soon some other, numerous

> villagers could "recover" sacks of corn and potatoes from the collective farm, thereby recouping some of what they had been obligated to contract from their private plots. This enabled them and their urban relatives to eat better than they "ought" to. It also enabled a few other urbanites to avoid standing in food lines in October for the winter's supply of potatoes because—using the extra gas they had bribed from the gas-station attendant—they would drive their cars directly to a village and pay five times the market price to buy forty kilograms of potatoes from some peasant. The practice naturally furthered urban food shortages and was one reason why policemen randomly stopped cars to spot-check for transport of food, which they would confiscate.[71]

Socialist ideology lauded the ability of the state to care for its people, prioritizing the group over the individual. In Romania, however, the state failed to care for its own, particularly those infants dying of AIDS or living with HIV and orphaned children. In 1980s Romania, the state-run care of many children—those with HIV or AIDS, orphans, the disabled, and those who had been abandoned—was horrifyingly nonexistent. It is estimated that during the Ceauşescu regime, as many as fifteen thousand children died in state care.[72] Not only were there more children in state care than the state had capacity for, but state-run institutions would profit off the payments received per child, essentially killing children by starvation or neglect. People were misled about the care the state was able to provide. Romanian families are tightly knit, and it was common for extended family members to care for grandchildren, cousins, nieces, and nephews. But food and consumable rationing under austerity made this difficult, if not impossible, for many. By the 1980s Romanians depended on the state for most necessities, which made surrendering one's child to it for temporary care not unusual. In some cases, families had to pay the state for the care. Most intended to bring their children home once they could care for them again, and "this 'pawn shop' strategy often succeeded; in other cases, parents were unable to locate their offspring. It was only after the fall of the regime that they—and the rest of the world—learned why."[73] Many children, especially physically and mentally disabled children, were left to die in state institutions because the state wrote them off as having no value as future socialist workers.

> Labeled "nonproductive" in a society ideologically dependent on production, the handicapped were effectively sentenced to death. "It is a place to go to die," said one medical director of a facility from which the productive and nonproductive were selected to be sent to other institutions, where they awaited their fate.[74]

The surge in HIV infections and AIDS, too, contributed to the death of an unknown number of infants and children in state care. Only in 1985 did Romania first publicly recognize HIV and AIDS, four years after most countries. A confluence of factors contributed to the unusually high number of

Romanian HIV-infected infants. Blood was never screened for the virus. Ceaușescu's ban on abortion and contraception, positioned as "pronatalist," made doctors potentially liable if an infant died in their care. To combat infant mortality, they commonly gave babies antibiotic injections, usually with reused, unsterilized needles as disposables did not exist and sterilization was not common practice.[75] The reuse of needles and lack of blood testing were a powder keg of HIV transmission in Romania, and children were disproportionately affected.

The Regime's End: The Most Violent of the Eastern European Revolutions

December 1989 marked the end of the decade, the communist era, and Elena and Nicolae Ceaușescu's lives. Romania was the last of the central and eastern European communist bloc countries to break from communism, and its departure was the most violent: more than one thousand people died and at least three thousand were injured in its revolution. Romania was also the only communist bloc country to execute its leaders: Nicolae and Elena Ceaușescu, after a brief trial, were executed on December 25, 1989, their end broadcast on Romanian television and across the world.

A confluence of factors, a few years in the making, contributed to those ends. In the latter half of the 1980s, the Soviet Union and the communist bloc were questioning their political, cultural, and economic health. People asked for freedom. In Romania, those actions were mostly private, but a few were public. In 1987 the disgraced communist leader Ion Iliescu (b. 1930)—who would later serve multiple terms as president of postcommunist Romania (1990–96 and 2000–2004)—publicly asked for the "unhampered spreading and circulation of knowledge."[76] In March 1989 six RCP retirees wrote an open letter to Ceaușescu, criticizing his regime. The authors of this "Letter of the Six" were former regime apparatchiks who had contributed significantly to its excesses and failures. In response to the letter, Ceaușescu placed all six under house arrest; none recanted, fanning Ceaușescu's rage while showing the limits of his power. In November 1989, a worker revolt broke out in the Transylvanian city of Brașov. Similar to what happened after the 1977 Jiu Valley miners' strike, momentum for dissent gained little traction. Also fueling discontent in the final decade were the daily struggles of life under austerity. Some believed that there was no end in sight, even though the external debt was largely paid off by 1989. Yet, the regime did nothing to remove or reduce the austerity policies.

In December 1989, the spark of revolution ignited in Romania's western city of Timișoara. A series of protests followed the regime's eviction of the Hungarian Reformed Church pastor László Tőkés from his post in the city. Tőkés was not new to speaking out about the regime's repressions. In 1986 authorities moved him from Transylvania to Timișoara to limit his influence among the ethnic Hungarian minority. In March 1989, after several months of unsuccessful intimidation, the regime suspended Tőkés and ordered his eviction from Timişoara. In December he lost his appeal and asked parishioners to gather on the eviction date outside his residence, adjacent to the church. On December 15, a group of three dozen, most elderly, parishioners gathered there, near a busy square and tram line stop. Others, curious, waiting for the tram, made the group appear larger. News spread about the "large gathering." The regime recognized but did little to quell that initial gathering. This gave people hope. In the days that followed, more joined: "By the next day, rumors were sweeping Timişoara that the Securitate and the city's authorities had been powerless in the face of open defiance and this, in turn, brought many more people to the church to see if the story was true."[77] Interestingly, it was in socialist workers' spaces like factories that people were able to gather, discuss what was happening, and plan what to do next. The crowd grew and so did their demands; their objective changed. People shouted slogans like "Down with Ceauşescu!" "Down with Tyranny!" and 'Liberty!' As Tőkés recalled, "The crowd forgot the initial reason for their resistance, and in general terms, they opposed the regime itself."[78] Violence followed. Protesters threw rocks, and police set fire hoses on them. Very early on December 17, authorities broke into Tőkés's home, made him sign a blank piece of paper to craft his acceptance of the eviction, and moved him and his wife back to Transylvania. The next day, when Tőkés was visibly absent from Sunday services, a group of about two thousand moved to the city's RCP headquarters, where security forces met them with fire hoses. Protestors threw stones and Molotov cocktails. Hand-to-hand combat broke out. The regime responded with lethal force, marking the first deaths of the revolution. Management at factories and other industrial sites met with workers to discuss the situation. Workers responded with complaints and started several small-scale strikes. People outside Timişoara learned about the events primarily through RFE, inspiring protests in cities across Romania.

Even though he knew of the events in Timişoara, Ceauşescu left Romania on December 18 for a planned state visit to Iran. He returned and addressed the public from the RCP headquarters' balcony in Bucharest. He had trouble getting the crowd's attention, starting to appeal, "Citizens of the capital …," then stepped away from the mic. Elena, at his side, encouraged him: "Talk to them, talk to them." To appease the crowd, he then loudly shared a laundry list of salary and social welfare benefits passed that morning. He ended by commenting on "the events of Timişoara," attributing them to "outsiders," clandestine forces with the objective to destroy Romanian "integrity, independence, and sovereignty," a paranoia he held that foreign agents were interfering in regime activities.[79]

Those from outside and inside the regime joined the revolutionary actions: "Large sections of the army, together with elements inside the RCP leadership, started to lay plans to remove Ceauşescu, while at the same time, the news of events in Timişoara provoked widespread popular revulsion, which paved the way for outright revolt."[80] On December 22, Ceauşescu tried to address the giant crowd outside the RCP headquarters. He was met with people's anger and the reality of revolution. He fled. A helicopter took him and Elena from the building, flew outside of Bucharest, and landed. They hijacked a driver and his car but were apprehended, held at a military base, interrogated, and subjected to a quick trial. In the days between Ceauşescu's last speech and his last breath, street protests turned violent and bloody. A new political party was formed, the National Salvation Front (Frontul Salvării Naţionale, FSN), which took over the state-run television station and began round-the-clock broadcasting of the events as they unfolded. On December 25, a firing squad executed Nicolae and Elena Ceauşescu, with their deaths filmed and shown that same evening on television in Romania and across the world.[81] The end of communism and the Ceauşescu regime in Romania had arrived.

Notes

1 Tismăneanu, Dobrincu, and Vasile, *Comisia Prezidenţială*, 175; Vatulescu, *Police Aesthetics*, 32, 36; Verdery, *Secrets and Truths*.
2 Judt, *Postwar*, 192.
3 Petrescu, "The Afterlife of the Securitate," 406.
4 Tismăneanu, *The Tragicomedy of Romanian Communism*, 42.
5 Robyn L. Myers, Illinois State Academy of Science, email message to author, September 20, 2016.
6 Siani-Davies, *The Romanian Revolution*, 22.
7 Roper, *Romania*, 52–55.
8 Ashley and Munteanu, "Ceauşescu's Leadership Position" 543.
9 Decret pentru reglementarea întreruperii cursului sarcinii, state decree no. 770, October 1, 1966; Kligman, *The Politics of Duplicity*.
10 Janos, *East Central Europe*, 300.
11 Tismăneanu, *Stalinism for All Seasons*, 197.
12 Siani-Davies, *The Romanian Revolution*, 24.
13 C. Popa, *Elena Ceauşescu*, 289.
14 Bachman, *Romania*, para. 2.
15 C. Popa, *Elena Ceauşescu*, 55–56.
16 Popa, 49.
17 Deletant, *Ceauşescu and the Securitate*, 338; Nagat, "Ceauşescu's War," 233.
18 "Romania Earthquake," Noel Bernard to Sig Mickelson, memo, n.d., RFE/RL corporate records, box 329, folder 15, HILA.
19 Decret privind regimul aparatelor de multiplicat, materialelor necesare reproducerii scrierilor şi al maşinilor de scris, state decree no. 98, March 30, 1983, articles 1–3.
20 Smaller ethnic groups include Ukrainian, Transylvanian Saxon, Swabian (or German), Lipovan, Turk, Crimean Tatar, Serb, and Slovak, followed by even fewer numbers of Bulgarians, Krasovani (ethnic Croatians), Greeks, Jews, Czechs, Poles, Italians, Chinese, and Armenians. D.3534, v. 4, 244, MI/DS-J, CNSAS.
21 Deletant, *Ceauşescu and the Securitate*, 305.
22 Turnock, "The Planning of Rural Settlement," 252.
23 Sampson, "Urbanization," 513.
24 Ceauşescu, *Report of the National Conference*, 58–59; Deletant, *Ceauşescu and the Securitate*, 301.
25 Hamilton, "Urbanization in Socialist Eastern Europe."
26 Chirot, "Social Change," 474.
27 Massino, "Something Old," 41.

28 Deletant, *Ceauşescu and the Securitate*, 308.
29 Deletant, 50.
30 Verdery, *Secrets and Truths*, 63.
31 Verdery, "Romania's Securitate Archive," 9–14.
32 Verdery, *Secrets and Truths*, 70.
33 Verdery, "Romania's Securitate Archive," 11.
34 D.123, v. 9, 67–77, D, CNSAS.
35 CNSAS, "Ministerul de Interne," 19–20.
36 A lieutenant colonel earned about 7,800 Romanian lei; see Deletant, *Ceauşescu and the Securitate*, 335; Immigration and Refugee Board of Canada, "Rank Hierarchy."
37 Verdery, *Secrets and Truths*, 129.
38 Most of the *Securitate* journals are available through the Consiliul Naţional pentru Studierea Arhivelor Securităţii (CNSAS) website, http://www.cnsas.ro/periodicul_securitatea.html. See also "Petrescu, "The Afterlife of the Securitate," 391, 392n18.
39 Verdery, *Secrets and Truths*, 207.
40 Deletant, *Ceauşescu and the Securitate*, 335.
41 Petrescu, "The Afterlife of the Securitate," 390. For documents that supported Ceauşescu's case, see CNSAS, "Ancheta P.C.R din 1968."
42 Petrescu, "The Afterlife of the Securitate."
43 Petrescu, 398.
44 Neagoe-Pleşa and Pleşa, *Securitatea*, xxvii.
45 Verdery, *Secrets and Truths*, 17.
46 Tismăneanu, Dobrincu, and Vasile, *Comisia Prezidenţială*, 382.
47 Stan, "Inside the Securitate Archives.".
48 Tismăneanu, Dobrincu, and Vasile, *Comisia Prezidenţială*, 385.
49 Verdery, *Secrets and Truths*, 97–98.
50 Tismăneanu, Dobrincu, and Vasile, *Comisia Prezidenţială*, 385–86.
51 Verdery, *Secrets and Truths*, 99.
52 Tismăneanu, Dobrincu, and Vasile, *Comisia Prezidenţială*, 392.
53 Vultur, "Daily Life," 431.
54 Dobre et al., *Acţiunea*; Lambru, "Currency Exchange Operations."
55 Banu,"Capitaliştii *avant la lettre*."
56 Rosca, "IKEA Funds."
57 Banu,"Capitaliştii *avant la lettre*"; Lynn, "Cigarettes Good as Gold"; Thurow, "In Romania, Smoking."
58 Stan, "Inside the Securitate Archives."
59 Capotescu, "Migrants into Humanitarians," 293.
60 Raab, *All Shook Up*.
61 Capotescu, "Migrants into Humanitarians," 295.
62 Goma, *Culoarea curcubeului '77*.
63 Neagoe-Pleşa and Pleşa, *Securitatea*, xv.
64 Deletant, *Ceauşescu and the Securitate*, 245.
65 Deletant, 246.
66 Deletant, 249.
67 Becker, interview, 189.
68 Siani-Davies, *The Romanian Revolution*, 33.
69 Verdery, *What Was Socialism*, 40.
70 Siani-Davies, *The Romanian Revolution*, 33.
71 Verdery, *What Was Socialism*, 52.
72 Despa, "Starvation, Abuse."
73 Kligman, *The Politics of Duplicity*, 86.
74 Kligman, 227.
75 Kligman, 222.

76 Pittaway, *Eastern Europe*, 197.
77 Siani-Davies, *The Romanian Revolution*, 59.
78 Radio Free Europe/Radio Liberty, "Romania," para. 1.
79 Ceaşusecu, "Last Speech," parts 1 and 2.
80 Pittaway, *Eastern Europe*, 197.
81 *Washington Post*, "Transcript of the Closed Trial."

2 Experiencing the Earthquake and the Initial Response

A giant full moon lit a clear night sky, and temperatures were above average for late winter all across Romania. At 9:22 p.m. on March 4, 1977, an earthquake began and lasted about one minute. Measuring 7.4 on the Richter scale, it was the second-strongest earthquake to hit Romania in the twentieth century.[1] The capital, Bucharest, experienced the overwhelming majority of its destruction—90 percent of the county's 1,578 deaths, 67 percent of the 11,321 people injured, and 90 percent of the industrial and residential damage—because of the city's high population density, seismically unsafe high-rise buildings, and proximity to the earthquake's epicenter.[2] The industrial town Zimnicea on the Bulgarian border, the small city Craiova 140 miles west of Bucharest, and the former capital Iaşi in the east also experienced significant damage. Outside Romania, almost two dozen countries felt the quake.

The earthquake struck on a Friday while people were out on the town. "There was a nightlife in Bucharest, the city was lit up; there were restaurants, bars open until the early morning … people in the streets."[3] Many settling in at home were watching the Bulgarian film *Sweet and Sour* on state-run television; after the earthquake, the joke circulated that "God got mad, slammed his fist on the table, and said, 'A Bulgarian film again?!' and, then came the earthquake."[4] Romania's small Jewish community was observing Shabbat and preparing for the Purim holiday. Many children slept while adults sat in restaurants and cafés or attended a movie, play, or concert. The earthquake struck while the Philharmonic played in the Radio Hall, the live recording capturing the chilling sounds as the audience experienced it.[5]

People felt the earthquake's horizontal movements for nineteen seconds, and then the stronger vertical movements for about forty seconds more. The quake felled bookcases and wardrobes, kitchenwares, and canned goods. Those who left their apartment buildings as the quake shook saw plaster cracking, floor tiles erupting, and walls undulating. People gathered outside in their building courtyards: "We looked at each other with no understanding of what had just happened. The sounds were diverse and at times deafening from items falling, cracking, or crashing, windows breaking, and people screaming."[6] A US Embassy officer remembered,

DOI: 10.4324/9781032632933-3

> I had a four-year-old and a baby of less than one year. The first thing I screamed to my wife was, "Oh, shit, it's an earthquake and it's a big one. You get the baby and I'll get Michele (our older child)." I pulled Michele out of her bed just as a huge chandelier fell right where she had been lying.[7]

An expat child heard a bubbling and then a rumbling and thought a truck had hit the house. As she felt her bed bouncing, her parents grabbed her and her baby sister and ran down the stairs; halfway down, the movement suddenly stopped.[8] Some noticed dogs howling, chickens cackling, and children acting strangely moments before the earthquake. One expatriate couple prepared to go out to a dinner party. When their Romanian caregiver arrived, even though their child knew and loved her, the child had a meltdown, one so intense that the parents took her with them, sending the caregiver home. Their building fully collapsed from the earthquake, and so this change in plans saved all four.[9]

A nine-year-old at the time remembered the shaking "for what felt like a long time, but I've read it was only a minute." She was playing dress-up with a friend sleeping over. Her dad was at a business dinner, her sister was with friends, and her mother was in the kitchen when they heard a very loud rumbling. All went to the window, expecting to see a truck convoy. Instead, the apartment, located on the top floor of an eight-story building, began to sway and shake. Once it stopped her mother left the window, rushed to the kitchen, caught the fridge as it was about to fall over, and pushed it back into place. The lights and natural gas went out. They went down the stairs, her mom instructing them to press their backs to the wall.

> We used to play with the railing because it would vibrate when you banged on it. As we walked down with our backs against the wall the railings were still vibrating so profoundly you would have hurt yourself if you touched them.

Once outside, they piled into the family's car. She recalled "throngs of people in the street. They were trying to open the door to get in our car. People were just frightened. They didn't know what to do." Driving through Bucharest to an established meeting point, she saw "a building that was like a giant dollhouse where I could see the wallpaper on each level on each floor."[10]

The earthquake damaged a handful of buildings in that same way (shearing off the facade or a single exterior wall), but most devastating were the thirty-two high-rise buildings of residential and commercial use that fully or partially collapsed. Many had shops, cafés, and restaurants filled with customers on the ground floor and residents in apartments above. The six nonresidential buildings that completely collapsed included a hotel, three office buildings, the government's "computing" center, and the university's chemistry department.[11] A "frightened Bucharest telephone operator" reported that she saw "several … ten- and twelve-story buildings, housing hundreds of people, leveled in the capital. I saw legs. I saw heads. I saw very many body parts."[12] A medical student recalled,

> When the earthquake happened, I was riding the trolley bus. I saw some trees moving, but I didn't realize it was that bad.... When I got home, I found my television set fallen over in my bed and ... I still didn't believe it was that bad and I rushed to catch a cab.... I asked to be taken to the Emergency Hospital. The cab driver told me we couldn't go through the city center because the Scala building [14 Rosetti Street] had collapsed. I thought he was making up stories. He chose to go on Ştefan cel Mare [Road], and when we reached ... a collapsed building I saw shocking scenes. The screams inside were the first to unsettle me. Some [victims] were using flashlights to signal that there were people.[13]

If the full collapse was not horrifying enough, fires broke out in several locations. People recall the night sky bright from the moon and the dozens of fires across the city. Two buildings caught fire immediately; fires burned in the rubble of about ten of the collapsed buildings. Some of these fires persisted for days.[14]

Almost all who died in Bucharest perished in the partially and fully collapsed residential buildings clustered in the capital's city center.[15] Many were built before 1940, a factor that contributed to their vulnerability. Within hours of the earthquake, the US ambassador reported to the US Department of State that "downtown Bucharest appears heavily damaged with a number of collapsed buildings and wide-spread structural damage. Have no estimate of casualties, but believe will be fairly heavy."[16] Dawn's daylight revealed the true extent of the loss and damage.

In Bucharest, first responders deployed, concentrating their efforts in the capital's center. A US Embassy Marine guard exclaimed,

> You can hear sirens running around. The people are all in the streets and the rescue teams are trying to do what they can in the darkness with all the lights out ... everyone is living in fear of a new tremor any moment and buildings are empty.[17]

First responders included all workers under the Interior Ministry umbrella: firefighters, police, army, and the Securitate. Interior Ministry workers responded immediately, many reporting to their posts before receiving an official call, and almost all were on the job that night.[18] The ministry "initiated the first measures to save victims and transport them to hospital and to remove people trapped under the damaged buildings."[19] Firefighters fought more than fifty fires burning at the same time, thirty-one in the capital and twenty outside of it.[20] They worked those first hours without street or interior lighting, telephone, or radio. Few had two-way radios, so most had no way to reach their commander. Several water lines broke, and, in some cases, firefighters used sewer water to extinguish blazes.[21] Tertiary damage of this sort is common following earthquakes: fires erupt from fireplaces and natural gas main breaks and are fueled by exposed flammables, while water to fight them is disrupted by water main breaks. The firefighters' mission went beyond putting out fires. They also saved people and transported victims. They aimed fire truck headlights for nighttime search and rescue as they helped dig through the rubble. They sprayed that same rubble with chlorine and lime to prevent victim remains from creating a health hazard. They brought water to factories, bakeries, and energy plants to keep production running.

Telephone lines were cut, and residents could not call for help. The collapsed buildings' rubble obstructed roads, and officials closed and rerouted many. People mostly walked or ran to fire and police stations to report where assistance was urgently needed.[22] The fires, chaos, and obstructed streets hampered first responders' ability to get to the collapsed building sites to search for survivors and transport the injured for care. Because transportation was affected and the city did not have enough rescue vehicles, many people carried victims. An apartment building about four blocks from one hospital fully collapsed, a doctor remembered, "in about five minutes, the [intersection on the hospital's southwest corner] was completely blocked. It was an inferno! The healthy carried the wounded in their arms. It was a wave of arms covered in human bodies."[23]

The number of dead and injured overwhelmed the health-care system. In Bucharest, more than 2,300 people were hospitalized, and more than 11,300 sought medical treatment.[24] A medical student recalled, "When I reached the Emergency Hospital, about twenty minutes after the earthquake, soldiers had already brought and left bodies in front of the hospital."[25] Death and injury from earthquakes are due mainly to structures crushing or burying people, as well as to masonry, glass, or household items falling and resulting in head injuries, broken bones, ruptured internal organs, or cuts and bruises.[26] First responders and ordinary citizens brought the injured and dead to the city's thirty-five hospitals, three field hospitals set up by firefighters and the Securitate, or the city's soccer stadium, where the Health Ministry established a triage center.[27] Within minutes of the earthquake, a couple hundred injured arrived at Bucharest's Urgent Care Hospital, overwhelming its ability to respond: the morgue rapidly reached capacity; the emergency room could not absorb the onslaught of injured, so care was triaged and rushed; beds filled quickly; personal protective equipment ran low; water was scarce; and staff had the additional task of calming traumatized loved ones. One medical student recalled, "We had nothing to drink; washing your hands was not even a possibility."[28] Not all hospitals were safe either. Officials closed several for fear they might collapse. At the Floreasca Hospital, an engineer quickly determined that it was at risk of collapse and ordered 1,500 people—health-care providers, staff, and patients—to evacuate. Patients there during the quake, if able, evacuated themselves to courtyards, joining doctors, nurses, staff, and the many injured and their families. One emergency room doctor recalled, "the wounded were screaming and crying and were left in front of the hospital. The courtyard was also filled with hospital patients who had come outside with blankets on their heads afraid the hospital, too, would collapse."[29]

Few health-care workers had experienced a disaster of that magnitude. A medical student recalled,

> our supervisors told us to deal with the situation, do what you can ... we stitched up wounds with no anesthetic and we didn't hear one groan. [The patients] were extremely courageous. The death of a young woman upset me a lot, she was brought nearly severed in two and she told me not to care for her because she would soon be gone anyway. And, so it was: terrible.[30]

Hospital staff helped patients in hallways, courtyards, and wherever they could, triaging as they did. A supervisor instructed medical students to

> go out of the hospital and get patients from [the courtyard] and check for a pulse. If they have a very good pulse, leave them alone. If they have a really weak pulse, leave them there because they will die, there's nothing we can do. If they have an intermediate pulse ... we'll see from there if we can save them.[31]

Writer Romulus Rusan, cofounder of the Memorial of the Victims of Communism and the Resistance (Memorialului Victimelor Comunismului şi al Rezistenţei), was among the victims with a strong pulse who waited for medical treatment. His apartment building partially collapsed; a bookcase and part of a wall trapped him for three hours. He recalled that

> after an hour I heard someone call, "Is there anyone here?" I waited for the neighbors to say something first because I heard them, but when they did not say anything, I called out. I saw light through the rubble, there was a full moon that night, but it was someone with a flashlight.

That someone, a police officer, eventually dug him out. Rusan walked to the hospital and waited for treatment. When a doctor finally came, he asked Rusan whether he smoked. He told the doctor he did not. The doctor asked again. He repeated that he did not. The doctor pointed to Rusan's foot, which was shod with an empty Kent cigarettes carton. Rusan had left his building shoeless and took the carton from a corpse near his partially collapsed building.[32] Cigarettes were an extremely valuable currency used in the secondary market, often as payment for extra or preferential services, like health care, with the American Kent brand the most desirable. Luck had found Rusan pinned under a bookcase, but not crushed. Luck, too, had meant that a first responder heard his cries and dug him out, allowing him to get to the hospital and receive medical care. And, he recovered fully from his injuries. Rusan's experience was full of serendipity, thousands were not.

More than 1,400 people died in Bucharest, and families stood in long lines at the city morgue to identify their loved ones.[33] When no family could, the regime asked neighbors to come to the morgue to help identify the dead and called in workers to photograph the corpses and body parts. Dealing with the dead was an overwhelming task, as one doctor recalled:

> After a time, they started bringing bodies to the [Forensic Medicine Institute] and it was a terrible sight. There were parts of human bodies laid out on sheets of plastic and people waiting at the gate, relatives to identify them.... It was a shock to us and we lived in the medical world ... you can imagine what it meant to go to a place to see dead bodies, body parts strewn there, to try and recognize a family member, a loved one.[34]

Families chose where to bury their loved ones. For those who died in the capital, Bucharest's city council covered most of the burial costs for the identified and all the costs for the unidentified, who were buried in separate graves, numbered and indexed by where the remains had been found. Also buried were unidentified body parts estimated to be from an additional fifty-two people.[35]

Food and potable water were made available. The government kept stores open longer and prohibited alcohol sales for the two days following the earthquake.[36] That first night the city was without light, water, or telephone, but within a day, the regime restored electricity, cold water, and phone service to residents, although it rationed cold water to six hours a day in two-hour blocks each morning, afternoon, and evening.[37] The government intentionally left the natural gas disconnected to prevent new fires, leaving residents for four days without hot water or the ability to cook food.[38]

Romanians helped first responders in the disaster recovery efforts, but only in ways the regime permitted. It discouraged citizen participation in the immediate recovery efforts and placed those firmly in the hands of the Interior Ministry. Bucharest residents helped as best they could. They brought food and hot tea to the workers digging through the rubble. Many people stood in the streets and watched. US Ambassador Harry G. Barnes Jr. sent a telex the morning after the earthquake, reporting that many stores were open. He observed

> lines of shoppers, clearly longer than usual, but we have seen no evidence of stampede and only a couple instances of hoarding. Radio has issued a continuous stream of announcements of diversion to Bucharest from surrounding cities of food supplies. Lines of cars at service stations are extremely long.

He continued that people

> have recovered quickly and thus far exercising admirable control.... We have been impressed with discipline and restraint exhibited by most Bucharest residents. After confusion and traffic jams of last evening,

surprising tranquility has returned to city. Lines in front of stores are orderly. Flocks of sightseers who have descended upon center of city appear more curious than distraught.[39]

The Initial Response

In the early morning hours after the earthquake, on an airplane returning to Bucharest, Ceauşescu enacted a state of emergency for five days across Romania and ten in Bucharest.[40] Its primary focus was not to search for survivors, care for the injured, or retrieve the dead, but rather to restore production at industrial sites. The economy's health was a primary objective of the Ceauşescu disaster recovery. Under the state of emergency, the government mobilized the Interior Ministry forces, "with all the necessary ... material and human resources at their disposal," as well as "all activists of the party and the state and of public and mass organizations" for the disaster recovery efforts, in particular to help "the population and the economic and social units affected by the earthquake."[41] "Activists" meant Romanian Communist Party (RCP) members and sympathizers. "Economic units" included transportation networks, factories, food processing plants, collective farms, mines, energy plants, and so forth. "Social units," in contrast, meant public institutions such as schools, hospitals, libraries, and cultural centers. The order in which the state of emergency listed ministries and agencies and their tasks reveals the regime's priority on repairing important economic entities. It outlined the restoration of telecommunications, transportation, and energy networks before the mobilization of health and sanitary workers or the provision of necessary water and food. It did not mention evacuations, temporary shelters, residential damage assessments and repairs, or long-term housing provisions. Article 9, however, required all "units to take immediate measures to save goods for the quick reestablishment of activities and the normal functioning of the means of production."[42]

During the state of emergency, the Interior Ministry prioritized first responders' deployment to locations that were vital to the state-run economy and that had been damaged or placed at risk of further damage, as Ceauşescu insisted that economic activity be restored to full capacity.[43] Firefighters worked practically nonstop the first two days following the earthquake and then in more manageable shifts for more than a week in Bucharest and fourteen counties.[44] They extinguished blazes at the easily accessible fires but prioritized those at oil drills and refineries, power plants, electricity transfer stations, medical clinics, a large bread factory, and research and education institutions. Firefighters worked for more than a week at some of the important economic locations, including chemical plants in three cities, three petrochemical plants in one city, and the shipping port in Constanţa on the Black Sea.[45] Ceauşescu demanded that workers at the undamaged industrial, agricultural, and other production assets "work intensely, in long shifts to supplement" the potential economic losses resulting from earthquake damage.[46]

The economic assets affected included power stations, transportation hubs, telecommunications centers, research and educational institutions, chemical- and petrol-industrial sites, hospitals and clinics, food production and processing factories, water reclamation facilities, factories, and public utilities such as electricity, natural gas, and water.[47] The Interior Ministry issued daily, sometimes twice daily, updates on the functionality of these economic assets. Firefighters and Securitate troops helped demolish buildings at industrial locations severely affected and repaired those at risk of collapsing.[48] Where and when they responded was determined, in large part, by the economic importance of the building at risk.

Within days of the earthquake, the regime received disaster assistance from abroad. The Interior Ministry organized the receipt, transport, storage, and distribution of it all.[49] It received US, western and eastern European, Turkish, Greek, Chinese, and Soviet aid via busses, vans, planes, trains, and automobiles, with donations including powdered milk, baby food, medicine, and medical equipment. Italy sent cookies and candy, Bulgaria sent canned fish, and Norway, Belgium, East Germany, Hungary, Poland, the French automobile company Renault, and two French citizens all donated ambulances.[50] A Swiss team arrived with search and rescue dogs. The USSR sent two shipments of medicine and blood, and eventually an entire factory to produce prefabricated concrete slabs for constructing apartments.[51] A week following the earthquake a group of US seismologists arrived to gauge the possibility of aftershocks and warned the US ambassador about that possibility, which he shared when interviewed on Romanian state television.[52] This caused panic and disorder among Romanians, who had heard nothing about aftershocks and proceeded to evacuate their homes.[53]

Housing for those displaced from the partially and fully collapsed buildings was another concern. Starting on the second day after the earthquake, the government discussed how to resettle displaced residents. Ceauşescu told Political Executive Committee (PEC) members, "People must be evacuated; if we give them apartments, then give them quickly; a person must also have furniture." Elena Ceauşescu noted that evacuees did not have anywhere to take their furniture, to which her husband stated, "In the first place if they want to take furniture they will take it; if not, they won't."[54] She explained that they planned to give victims "furniture, kitchen items, six knives for each. It was assigned to each family—unfortunately, we only have a few families—many people are the only ones left [in their family]."[55]

The government prioritized resettling the collapsed building survivors. Within two weeks, half received ration tickets to buy new clothing, but very few received cash, or were placed in new, furnished apartments or shelters.[56] The regime proposed allowing them to purchase one- or two-bedroom apartments as home ownership was not uncommon. Nowhere did the regime indicate the amount, if any, of monetary assistance that was to be provided to them.[57] Four months after the earthquake, Bucharest officials reported that

151 families remained in dormitories as they waited for their homes' structural repairs to finish—or even start.[58] This small number, not even half of 1 percent of the more than 35,000 families displaced, indicates that the regime did little to help. Most found shelter on their own, usually with family. For those who were evacuated from damaged (rather than collapsed) buildings, the government did even less. It did not give cash for resettlement or free clothing to these people, at least half of whom went to live with relatives.

Atypical Responses: Securitate Troops as First Responders and Radio Free Europe's Open Channel

Two of the 1977 earthquake responses were atypical vis-à-vis practices in other countries. First, the Ceauşescu regime deployed Securitate troops to work as first responders. This was not a new practice for Romania. A few years earlier, the regime had deployed the Securitate as part of its disaster response to severe flooding. The second unusual response was an uncoordinated, yet joint effort, between the Western-supported, Munich-based Radio Free Europe (RFE) and the Ceauşescu regime. RFE opened a communication channel previously unavailable not only to Romanians but to all those across the communist bloc when its Romanian desk began transmitting live informative programming and airing telephone calls from Romanians outside and inside the country as part of its earthquake coverage. While it was illegal to listen to RFE, the Ceauşescu regime allowed the calls from Romania to be patched through to the foreign radio station.

Securitate Troops as First Responders

From the first moments after the earthquake struck, Securitate troops worked alongside firefighters and police. The regime mobilized all human resources for search and rescue, survivor assistance, economic asset protection, cleanup, and other disaster recovery efforts.[59] To do so it turned to the Securitate. While unusual outside the communist bloc, this practice was not new for Romania. Earlier in the decade, in 1970 and 1975, the regime had mobilized its Securitate to respond to severe flooding.

During the 1977 earthquake response efforts, the recently reorganized Interior Ministry supervised all Securitate, firefighter, and police troops. The ministry had an overall mission of

> protecting the state security, socialist, and personal properties, and people's rights and interests, preventing and discovering infractions, ensuring public order, and controlling international borders ... undertaken with the collaboration of the Securitate and [police] units and their subordinates all to ensure the peaceful work of socialist construction of the Romanian people.[60]

On the first day after the earthquake, all across Romania, the Interior Ministry deployed more than twice as many Securitate officers and troops as firefighters.[61] In Bucharest, the disparity was even greater. During the ten days following the earthquake, Interior Ministry commanders supervised two to four times more Securitate troops than firefighters in disaster response activities.[62]

It would have been difficult to identify the Securitate troops because they were disguised as "workers, intellectuals, or simply police." Their objective: calm panic and fight rumors through population control and information collection.[63] They also dug out and transported survivors and victims, and they guarded the collapsed building sites, factories, museums, archives, and shops in the capital and across Romania.[64] For the displaced, they secured temporary housing. They helped pick through the rubble for valuables, which they inventoried, transported, and stored. They secured arms, munitions, explosives, and archival documents.[65] For the regime, they tried to guide journalists', diplomats', and others' perceptions of the disaster response. They accounted for the dead and injured and tracked missing residents from the collapsed buildings.

One way the Securitate collected information was by tracking people, yet it found this task difficult during the disaster response. Securitate officials reported that "the difficulties in knowing precisely what happened to every resident of the buildings affected by the earthquake led to the need to create a population accounting system" to track "each and every resident."[66] The Securitate treated the 1977 earthquake disaster response as it did individuals: it was a target, or object, to follow. It was a contained event with tangible physical results that the Interior Ministry workforce responded to within a disciplined framework, activating established protocols. During the state of emergency, officers updated daily, sometimes twice daily, tables of the dead, injured, hospitalized, and housed individuals. They tried to account for every resident of the collapsed buildings, listing identified and numbering unidentified victims, as well as those hospitalized, rehoused, and missing (Appendix B). During the earthquake disaster response, Securitate officials also worked as bureaucrats and functionaries, documenting what they thought was important or what was just plainly available to document. They compiled, saved, bound, and archived documents created during and about the disaster response, titling the collection "The 1977 Earthquake," essentially creating the earthquake's Securitate file.[67] It must be stressed that the Securitate's primary disaster response objectives were to maintain public order and safety and prevent activities the regime characterized as undesirable. As its officers worked alongside firefighters and police, they also gathered information from people lingering around the damaged and collapsed building sites and deterred others from doing so. In and outside the capital, Securitate troops detained people they perceived to be potential "enemies of the state," preventing them from traveling to the collapsed building sites and other areas. Specifically, they identified and deterred people with criminal records from traveling. During the state of emergency, the Securitate prohibited "a large number of people with criminal records" from traveling to damaged areas in the capital and across Romania as they were thought to be "predisposed to commit crimes."[68] Together, these actions accomplished one of Ceauşescu's primary objectives of the disaster response, what he described as the need to eliminate "chaos" and "panic." It was common practice for the Securitate to report on the population's mood (*starea de spirit*), which they did following the earthquake as well. From the hours and days following the earthquake, through the months until it stopped assessments and repairs in July, the regime tracked any real, imagined, or fabricated cases of "chaos" and "panic" exhibited by people concerned about their damaged homes and businesses.

As clear evidence of the success of using Securitate troops as first responders, the regime amended its first responder protocols and asset protection plan two months after the earthquake, increasing communication channels and prescribing tasks for firefighters and police similar to those of the Securitate.[69] The objective was to prevent potential damage by securing industrial assets crucial to the state-run economy. Specifically, the new guidelines gave first responders the power to "discover and liquidate acts of diversion or sabotage that might undermine the national economy" and to defend "socialist property" against fires, explosions, or damage by "offenders" or others who attack public property. The amended protocols expanded the roles and activities of Securitate troops and those in its network to work in the capacity of first responders.[70]

Radio Free Europe's Unprecedented Earthquake Coverage

In Munich, minutes after the earthquake, RFE's Romanian desk lost contact with Bucharest. During the first hour and a half after the earthquake, Romanian state-run radio was out, either cut by earthquake damage or intentionally silenced by the regime. RFE's Romanian desk staff remembered,

> We knew from our Listening Section that Radio Bucharest was no longer broadcasting and the conclusion was immediately that something really serious had happened. But still, given that no information of any kind came from Bucharest during that period of several hours, no one could imagine how serious that earthquake really was.[71]

RFE staffers contacted the US National Earthquake Information Center, confirming the earthquake's strength, and broadcast news of it an hour and a half before Romanian state-run radio.[72] RFE director Noël Bernard interrupted the regular program and began unprecedented nonstop live programming.

Bernard recalled that in those first hours after the earthquake, "we did not go off the air at midnight, as usual, but continued non-stop, with great demands upon our staffers, until Sunday at midnight."[73] RFE shared any and all information it received, relaying

> where you could find a pharmacy, where you could find this, that, and the other. So, we took those immediately as they came and broadcast them back. Whatever advice we had at that time we broadcast back and we kept it up all night with these announcements and with music and filling in with whatever wire service stuff we had on the subject until the early hours of the morning.[74]

When no information was available, it

> played classical music between news announcements, broadcast [state-run] Radio Bucharest communique, used reaction from correspondents in Paris, Rome, Washington, and New York, gave Western press play, provided a history of earthquakes and scientific commentary, and reported on meetings and other developments in Bucharest relating to the earthquake and measures to cope with the disaster.[75]

State-run Radio Bucharest came back on air less than three hours after the earthquake, but RFE continued its nonstop broadcasting for almost a week. It also began broadcasting direct telephone calls, creating a bridge for communication between Romanians abroad and those in the country. An engineering expert who called to explain technical matters inspired RFE to patch calls through and air them live. Bernard explained that

> an engineer, a Romanian construction engineer, called me from Berlin and was giving me advice about what we should say on the air about repairs, and I said, "Look, I'm not an engineer and I don't have time to write it down … why don't you say it?" [He refused, saying] 'no, I can't, I have relatives in Romania, I'm traveling there.'[76]

The engineer avoided going on air for fear of retribution from the Ceaușescu regime, but the next day he called the station back. Bernard explained,

> we had two engineers and two architects, Romanians who were working in Munich, who came to a roundtable to discuss the whole thing and during that roundtable, this man from Berlin called again and he said, "Look, I now want to say my peace," and he got on the air and said some very reasonable and constructive things. And that broke the ice, in other words, although we have a very large [Romanian] audience in Germany, these people generally don't get in touch with us because they want to go back to visit their friends and relatives, they want to go back to Romania, and they feel that if they get in touch with us that might hurt their prospects there. Ah, so I invited others to call with advice, with messages, whatever. And the calls started flowing in from Romanians in Germany and other places.[77]

Following the Berlin engineer, the next on-air callers were from the diaspora.

> The plain black telephone in studio 13 at English Garden in Munich was ringing all weekend. "Melbourne here … Tehran calling … This is Helsinki." From around the world, Romanians were calling Radio Free Europe's Munich programming headquarters to ask if their relatives were safe after last week's massive earthquake. RFE, the US-financed shortwave broadcaster to Eastern Europe, scrapped regular programming on its Romanian service to beam calls live into Romania around the clock.[78]

Romanians abroad called RFE and asked about their loved ones, citing specific addresses, and Romanian listeners relayed those requests. To get answers, Romanians traveled by car, moped, bicycle, and foot to those addresses and phoned the station back with updates.[79] RFE later also began including phone numbers in the on-air calls, an idea that came from a group of locals who could not access addresses when driving around Bucharest because many streets were closed.[80] They formed a telephone tree. In one case,

> friends organized a listening group, taking down phone numbers of people in Romania as broadcast by their relatives in the West, and each of them put through approximately sixty calls informing those people about the broadcast and asking them to get in touch with their relatives abroad.[81]

As a result, some people received hundreds of calls.

> I remember making a lot of phone calls to ask everybody if they were alright. And among the last of them, as I was certain that nothing could've happened, I reached a desperate mother, her son had died while visiting someone in the Dunărea building [3–5 Bălcescu Boulevard]. I froze in horror.[82]

During the first twenty-four hours after the earthquake, calls came exclusively from Romanians abroad. In the days that followed, those in Romania reported back with information. Within days, calls about and for information increased as more people learned about the station's live on-air programming. Bernard recalled, "Immediately the next 24 hours it was almost exclusively calls [from Romania] answering people in the West, people who said, 'I'm well,' and thanking us for what we were doing." Telephone operators connected about six hundred calls to the station's live broadcast, as many as one hundred in one day, announcing as they patched in direct to the studio, "Hello Radio Free Europe, I have another call for you."[83] Often staying on the line, some operators spoke up during the live broadcast to clarify what the caller had said. For many

Romanians, international calls were arduous, personally risky, and expensive. At first, the regime took little notice, allowed calls to go through, and did not jam the airwaves, either directly permitting or simply overlooking the broadcasts. The Washington, DC, director of Radio Free Europe/Radio Liberty observed that RFE's Romanian desk

> provided not only a public service to the victims of the quake and their relatives and friends but also acted as a surrogate press in providing citizens in all the Eastern listening areas with speedy, accurate reports that often were not available in domestic media.[84]

In a typical disaster response, people usually have access to information from government authorities. This was not the case in Romania during the Ceauşescu regime, and so Romanians tuned to RFE following the earthquake for information. This, too, represents a practice that was not new in Romania at the time. By 1977, more than 50 percent of Romanians listened to RFE and represented its argest audience in central and eastern Europe.[85]

Two days after the earthquake RFE ended its nonstop programming, but it continued with the phone-in live programming for another ten days.[86] As the days passed, fewer people called to give or ask for information about traffic, utility service, and transportation, and more went on air to thank the station: "I am Tanase G. Ioan from Găeşti, 80 Independenţii Street in Dâmboviţa County, phone 10-73-7. I have a personal message for Noel Bernard.... I want to express my thanks, gratitude, respect, and consideration."[87] Romanians were grateful for the information and the open platform RFE provided.

The regime allowed telephone operators to connect many of those calls, but it also cut many of them off. It was the US-based RFE directors who ended the live call-in programming, out of concern that Romanians were using it to criticize the Ceauşescu regime. RFE directors wrote to Munich staff:

> In view of changing nature of calls received by Romanian [broadcasting desk] during its periods of public service message broadcasting ... believe serious consideration should be given to reducing or eliminating this service earlier than is suggested.... The worst and most critical period of the earthquake emergency is already past, and RFE is not meant to substitute for the use of international telephone and cable communications facilities.

RFE staff acknowledged and shared with the DC home office that "more and more oddball cases may be cropping up."[88] Bernard remembered,

> we got a call from someone who claimed he was in Andorra who immediately began saying, 'this criminal Ceauşescu, since he came to power, we have floods, we have an earthquake, we have so on and so.' Well, he

> managed to say this before I could stop him and my answer was 'thank you very much, but I don't think you can blame Mr. Ceauşescu for floods and earthquakes.' That was it. There was another one who called from Romania and began, 'Ceauşescu is the Idi Amin,' she meant of Romania and was cut off at the other end before I was able to cut her off.[89]

Why the regime initially allowed and then cut calls is unknown and may have no rationale. Maybe Securitate staff on duty were sympathetic to the need for information, or their capacity to listen was diminished by the other pressing disaster response activities. Later that year, the regime began registering all international calls made from Romania, likely in response to Romanians' calls to RFE.[90] It was unprecedented for RFE to be the conduit between communist Romania and the West. When state-run radio returned, it failed to provide the public with adequate or actionable information. Only on the third day after the earthquake did the state-run national daily newspaper *Scînteia* publish any earthquake news. RFE's earthquake coverage was an exceptional example of the diaspora's and local Romanians' ability to penetrate the impermeable border the regime sought. Through radio, people crossed borders spanning thousands of kilometers, shared their concerns and information, and consoled one another. There was a need for objective and timely information, and RFE's open airwaves made space for that. As a result, RFE gained even more credibility, sympathy, and confidence.[91] The 1977 Romanian earthquake coverage was the first time it went to a live on-air format, something the Romanian broadcasting desk would return to during the revolutionary and regime-ending events of December 1989.

The Ceauşescus' Criticism of the Initial Disaster Response

When the earthquake struck, Nicolae and Elena Ceauşescu were in Nigeria, part of a twelve-day state visit to West Africa. Amid an evening reception in their honor, the ruling couple learned of the earthquake almost immediately after it struck. Ceauşescu's secretary was on the telephone with his wife in Bucharest when he heard her shout, "There's an earthquake! Get the child out of the house quickly!" Then the line was cut. Because phones were down, regime officials contacted the newswire service AGERPRES to be briefed. Elena immediately asked whether her home was damaged (it was fine). Nicolae sent a message to PEC members: "Do what you think must be done! I'm leaving tonight for home."[92] That night, Ceauşescu and his entourage flew back to Bucharest.

At the news, the ruling couple were at first perplexed and then agitated, concerned that regime apparatchiks might use this opportunity to seize power. Their fear was overblown. Leaders were paralyzed. That evening in Bucharest officials met at the RCP headquarters. Prime minister Manea Mănescu was "completely overcome," and the head of Ceauşescu's chancellery, Ştefan

Bârlea, was "unable to talk on the phone, gripped by hysteria."[93] Most regime leaders in Romania knew what they should do, but did little because, "who was brave or insane or stupid enough to take the reins, make decisions, impose himself in the position of leader? That would be political and possibly literal suicide. So those in Bucharest did nothing." They kept the state-run radio off air those first few hours, even though the radio transmitter was working. "No one dared to decide to restart the broadcasts, because no one knew what had to be said and what could be said. Romania was once again cut off from the rest of the world."[94] PEC members in Bucharest

> proposed creating an ad hoc committee to take care of the situation until Ceauşescu return[ed].... The proposal fell, however, at the moment when the question was raised: how do you mean, even just for a few hours, the country should be led by someone other than Ceauşescu? And of course, no one had the courage to take such responsibility.[95]

Some Romanians lamented that the ruling couple did not experience the earthquake: "Ceauşescu didn't even feel the quake, the leader being away at that time on one of his endless trips that were forbidden to the common citizen."[96] The Ceauşescus arrived early the next morning and immediately toured Bucharest.[97] Romanians noted their walking tour of the capital the morning after the earthquake, many impressed with the ruling couple's concern for the survivors, those injured, and the city's destruction (see Figure 2.1). After their tour, Nicolae and Elena Ceauşescu, along with Mănescu, Bucharest mayor Ion Dincă, and minister of the interior Theodor Coman, discussed the earthquake recovery efforts with the ministers of national defense, domestic commerce, labor, forestry, and construction materials; the first vice president of the Ministry Council; seven government vice prime ministers; the president of the State Committee on Nuclear Energy; the head of the state-run media; a representative from the External Relations Section of the Central Committee of the Romanian Communist Party (CCRCP); and the RCP Red Cross liaison.[98] During the first few meetings of the PEC after the earthquake, the Ceauşescus expressed great concern for the earthquake's survivors. After he saw firsthand the destruction and recovery work, Nicolae Ceauşescu stated, "in the first place, there must be actions to save people."[99] Then he criticized the Interior Ministry's efforts: "We must think about those who suffered damage, those who lost their furniture, clothing. We must assist them, but first, we must see. We will house them, but also give them some help." He asked whether there was enough food for the city's population.[100] Elena interrupted, correcting him: "First we must remove people from under the rubble."[101] The Ceauşescus' initial responses were not, however, all heartfelt and benevolent. Few would willingly describe Nicolae and Elena Ceauşescu as compassionate leaders. Alongside the instances of expressed compassion for the earthquake survivors and victims emerged glimpses of the paternalistic dictator and his

Figure 2.1 Nicolae (far right) and Elena Ceauşescu pass the collapsed Nestor building (2 Pictor Grigorescu Street, Stair A) as they tour Bucharest with other regime officials, March 5, 1977.

Source: Fototeca online a comunismului românesc, LA419, Cota 37/1977.

cruel wife. Just moments after asking for increased search and rescue efforts, Nicolae Ceauşescu said that "if some [victims] remain without clothing, that is fine."[102]

First responders continued the search and rescue for days, and Coman shared that the work was going well. The US ambassador reported to Washington that "the response in cleaning up most seriously affected regions and controlling movement appears reasonably prompt and effective," adding that the regime had

> quickly asserted control. Clean-up operations were started early this morning to remove the wreckage of collapsed apartments near the embassy. In the city center, these efforts have been hampered by throngs of passers-by, but a large number of police and soldiers are on scene.[103]

Three days following the earthquake, Nicolae Ceauşescu believed they could still find survivors, saying, "Concerning the work to be done, understand that under the rubble they still find people and some are still alive."[104] It is difficult for anyone to survive for more than twelve hours buried under rubble, however, and so state-run newspapers ran page-long features of those who did.[105] Ioana

Milanovici and her husband survived for almost two days inside a pocket of air under the rubble of their collapsed apartment building.[106] Fifty-eight-year-old teacher Elena Enache reportedly survived almost eight days, trapped without food or water.[107] Nineteen-year-old Sorin Crainic survived ten days trapped under the rubble, a story some believe was falsified (others believe that Crainic survived because Elena Ceauşescu insisted that the search for survivors continue when others in the regime thought it was futile). The grim reality, however, is that these stories were the exception; 1,420 people died in Bucharest from the full or partial collapse of the high-rise buildings.[108]

Those first days Nicolae and Elena Ceauşescu met with members of the PEC once or twice daily to discuss the recovery efforts across Romania, and in Bucharest in particular. At the first recorded meeting after the earthquake, Ceauşescu expressed his concern for Romanians' safety and outlined the regime's response priorities.[109] He rambled somewhat incoherently when expressing his initial concerns about the recovery efforts, consideration for the victims, and care for survivors, saying, "but working with housing and with these funerals, that is very many people. Therefore, we must assist them; we will take all measures to assist."[110] The US Department of State reported internally that "Ceauşescu, who was on the scene within 18 hours, lost no time in publicly criticizing certain responsible factors for shortcomings in the organization of urgent actions to eliminate the effects of the earthquake."[111]

Ceauşescu initially criticized many working on the disaster response, including first responders. The second day after the earthquake, during a PEC meeting, he commented first on the bravery of the first responders, then criticized Minister Coman for not demanding that his workforce enter partially collapsed buildings to search for victims. He complained that he had seen workers at the damaged building sites standing around doing nothing and learned that they did not want to dig into the rubble for fear of triggering a collapse. He claimed that workers

> are afraid that the building will fall on their heads. It's the same at the other café [in a partially collapsed building] … I went in with them and asked them where [they should] make a hole [to enter from the side]. I found an engineer, and he said that it is for the minister [to determine] and that he does not dare to enter. All stand and watch.[112]

He characterized the Interior Ministry workforce as cowardly, saying,

> there exists an unallowable mentality. Yesterday a way should have been found to enter [the rubble under the partially collapsed building]…. Some [workers] were afraid to enter in case [the building] crashed in on them. For us to speak of a spirit of sacrifice![113]

Ceauşescu blamed Minister Coman and Mayor Dincă for what he considered to be the disorganized recovery work on the first day after the earthquake.[114]

Coman reported that the night before, first responders had saved a three-year-old child, and he defended himself and his workers, stating that they were trying to find and remove survivors and had worked all night under difficult conditions, including having to light worksites with truck headlights. He explained, "we are working with bulldozers and when we see a body, we make a circle around it and start removing the rubble by hand, taking the bricks one by one, and so it takes a long time."[115] Ceauşescu responded,

> What am I to say? Not to use bulldozers? Over there where there was a café, a restaurant, and a large crowd, one must go down from the building next door, even under the rubble. Solutions must be found and not to wait until all the debris has been taken away. The problem is that others have not found a solution to enter through [the rubble]. Not all [solutions] are being tapped.[116]

Ceauşescu told Coman that there should have been a commander at each of the collapsed building sites to manage operations. Instead of describing the first responders in charge, Coman reported that he had successfully deployed political operatives—meaning RCP members, Securitate troops, informants, and others—explaining that by 3:00 p.m. the day after the earthquake, "it was stabilized everywhere in the city that there is a party activist." Ceauşescu exploded:

> Activists are something else. You are military and executives. What sort of exceptional state is this? The activist is set to mobilize politically.... Whoever is afraid he might be killed must be taken away. I have said since yesterday to be a commander, to be a boss.[117]

Ceauşescu saw Coman as unable to adequately mobilize the Interior Ministry workforce, encouraging its staff to be courageous, recover victims, and find survivors. He then asked why a temporary shelter he visited the evening after the earthquake was empty. When told it still had not been prepared to receive the displaced, Ceauşescu chastised his ministers and their ability to arrange housing:

> People are standing in the street with their children. We decided what to do and to get it done and still, we have resolved nothing. We are not in a state that in one day we will solve this problem! I said that if you don't have anywhere to house people put them in hotels. I told you there is a need to use the hotels because people are in the street. This is not allowable![118]

Ceauşescu was right. That first night, many in the capital were afraid of potential aftershocks and went out to the streets, sleeping in the open or in their cars. The earthquake's damage displaced more than 120,000 people.[119] During the state of emergency, the Interior Ministry provided housing, clothing, and food for 3,500 evacuees and temporary housing in more than two dozen different locations in Bucharest.[120] Officials evacuated some families to temporary housing in an exhibition hall, university dormitories, schools, hotels, and unoccupied apartments.[121] The regime resettled many victims discharged from hospitals, dormitories, or other temporary shelter sites. It placed some homeless in new apartments and published photo ops in the state-run newspaper. Very few people were displaced for long—but the regime was not solely responsible here. The regime resettled many of the collapsed buildings' few survivors, but the overwhelming majority of the displaced found shelter with relatives and friends.[122]

Families were key to the disaster response. By the 1970s Nicolae and Elena Ceauşescu had declared themselves the father and mother of the nation, building on Romania's cultural trait of extremely close family ties. The regime exploited that when it encouraged the majority of those displaced to find shelter on their own. RFE's Bernard later recalled the importance of family ties in the aftermath of the earthquake, describing initial calls to the radio station:

> from all over the place, from five continents, and practically all countries in Europe with messages for relatives, "please let us know how you are." It was at that stage that I realized something I had not been aware of before: how close and how deep family ties are in Romania. You know I lived in England for many years and in England you hear on the BBC an announcement every day, "Would so and so, who has not been heard of in thirty years, get in touch with his father who is dying," ah, this sort of thing is unthinkable in Romania; there are extraordinarily close and deep family ties in Romania and this catastrophe had made those rise to the surface and all other considerations were swept aside.[123]

By the late 1970s housing in Romania was already scarce from the urban planning policy, systematization, which required people to move from rural areas to either urban ones or urbanized rural areas. Bucharest in particular was densely settled, and new housing was scarce. Almost half of the capital's homes housed married couples with children, and an additional quarter had married couples with kids as well as other relatives sharing an apartment. The earthquake's damage to homes and subsequent displacement of people only exacerbated those already existing housing shortages and cramped apartment living, with the reliance on families for housing.

Conclusion

The initial disaster response under Nicolae Ceauşescu's regime included both typical and atypical activities. As most governments do, the Ceauşescu regime initially helped survivors, recovered victims, and managed assistance. In those first days, however, two unusual disaster responses also occurred, with

Securitate forces working as first responders and Munich-based RFE filling the regime's information blackout (a common practice by communist bloc countries in response to disaster).[124] In the months that followed, atypical disaster responses continued, with the regime prioritizing the economy ahead of helping survivors; dispatching Securitate troops to protect, collect, and inventory recovered valuables; nominally suggesting (but in practice mandating) labor and cash "donations" in the name of disaster recovery; organizing a clandestine operation abroad to influence the kind of humanitarian aid sent from the West; and finally, just four months after the earthquake, secretly ending required repairs, leaving hundreds of buildings housing tens of thousands of people unsafe to this day. The March 4, 1977, earthquake, as one survivor recalled, "was a nightmare that I would not want to live again or have anyone else live ever."[125]

Notes

1 Georgescu and Pomonis, "Building Damage."
2 INCD pentru Fizica Pământului, "Cutremurul vrâncean din 4 martie 1977."
3 Massino, "From Black Caviar," 236.
4 AGERPRES, "March 4, 1977."
5 See INCD pentru Fizica Pământului, "Cutremurul vrâncean din 4 martie 1977."
6 Alex, "Marturii de la cutremurul," posts 2 and 17.
7 Becker, interview, 186.
8 Laura Raicovich, pers. comm., August 30, 2023.
9 Laura Raicovich, pers. comm., August 30, 2023.
10 Robin Bacsfalvi, pers. comm., March 3, 2016.
11 D.11.737, v. 105, 168, DB, CNSAS.
12 *New York Times*, "Hundreds Reported Killed.".
13 AGERPRES, "March 4, 1977".
14 D.12.639, v. 15, 390 rev., DB, CNSAS.
15 Georgescu, Pomonis, and Steinbrueck, "New Archival Evidence," 3.
16 ActionAid Agency for International Development, "Earthquake: Sitrep 4 March 8."
17 This quote was widely circulated on March 5, 1977, appearing in the *New York Times*, *Baltimore Sun*, *Chicago Tribune*, *Washington Post*, and *Los Angeles Times*.
18 D.3635, v. 2, 122–22 rev., MI/DS-J, CNSAS.
19 D.11.487, v. 10, 229, DB, CNSAS.
20 D.12.639, v. 15, 390, DB, CNSAS.
21 *Adevarul*, "Fostul şef al pompierilor."
22 D.12.639, v. 15, 390, DB, CNSAS.
23 MediaFax, "Mărturii din 4 martie 1977."
24 D.11.737, v. 105, 141, DB, CNSAS.
25 AGERPRES, "March 4, 1977."
26 Hewitt, *Regions of Risk*, 206.
27 D.3635, v. 2, 125 rev., MI/DS-J, CNSAS; D.11.487, v. 10, 38–39, DB, CNSAS; Berg, "Earthquake in Romania," 5.
28 AGERPRES, "March 4, 1977."
29 AGERPRES.
30 AGERPRES.
31 AGERPRES.

32 Romulus Rusan, pers. comm., June 19, 2014.
33 Samuel Lyon to Ralph Walter, memo, March 11, 1977, RFE/RL Romanian Broadcasting Department records, miscellany 1977–79, HILA.
34 AGERPRES, "March 4, 1977."
35 Most victims were buried in Bucharest's Domneşti Cemetery. The total burial cost to the regime was 1,490,828 lei ($74,500). Secţia Cancelarie, 48/1977, 97 rev., FCCPCR, ANR; Secţia Administrativ-Politică, 12/1977, 51 rev., FCCPCR, ANR.
36 Secţia Cancelarie, 27/1977, 25, FCCPCR, ANR.
37 ActionAid Agency for International Development, "Earthquake: Sitrep 4 March 8."
38 Secţia Cancelarie, 20/1977, 12 rev., FCCPCR, ANR.
39 ActionAid Agency for International Development, "Earthquake Sitrep."
40 Decret 73/1977, 1–2, CCSPSR, ANR; Decret 74/1977, CCSPSR, ANR; Decret 75/1977, CCSPSR, ANR.
41 Decret 58/1977, articles 2 and 3, CCSPSR, ANR; Decret 73/1977, 1, CCSPSR, ANR.
42 Decret 73/1977, 1–2, CCSPSR, ANR.
43 Secţia Cancelarie, 18/1977, 13 rev., FCCPCR, ANR.
44 D.12.639, v. 15, 370, DB, CNSAS.
45 D.11.487, v. 4, 119 rev. and 385, DB, CNSAS.
46 Secţia Cancelarie, 18/1977, 13 rev., FCCPCR, ANR.
47 D.11.487, v. 4, 119 rev., DB, CNSAS; D.11.487, v. 10, 228–29, DB, CNSAS.
48 D.11.487, v. 4, 120, DB, CNSAS.
49 D.3635, v. 1, MI/DS-J, CNSAS.
50 D.12.639, v. 15, 310–10 rev. and 320–22, DB, CNSAS.
51 Secţia Cancelarie, 48/1977, 98 rev., FCCPCR, ANR; US Congress, "Romanian Earthquake Relief," 43.
52 Berg et al., *Earthquake in Romania.*
53 Joel Levy, pers. comm., July 11, 2023.
54 Secţia Cancelarie, 20/1977, 6 rev. and 7, FCCPCR, ANR.
55 Secţia Administrativ-Politică, 9/1977, 4, FCCPCR, ANR.
56 D.11.737, v. 105, 137–37 rev., DB, CNSAS.
57 Secţia Cancelarie, 35/1977, 93 rev., FCCPCR, ANR.
58 Secţia Economică, 78/1977, 3 rev., FCCPCR, ANR.
59 D.11.487, v. 10, 39, DB, CNSAS.
60 D.123, v. 9, 67, D, CNSAS.
61 D.11.487, v. 4, 120 rev., DB, CNSAS; D.11.487, v. 10, 227, DB, CNSAS.
62 D.14.800, v. 1, 4–5, DB, CNSAS.
63 Vlaicu, "În orele grele."
64 D.11.487, v. 10, 229, DB, CNSAS.
65 Vlaicu, "În orele grele."
66 D.3635, v. 1, 127, MI/DS-J, CNSAS; D.11.487, v. 10, 41, DB, CNSAS.
67 D.11.737, v. 104 and v. 105, DB, CNSAS.
68 D.11.487, v. 10, 37, DB, CNSAS.
69 D.3634, v. 6, 319–22, MI/DS-J, CNSAS; D.3635, v. 2, 105, MI/DS-J, CNSAS.
70 D.3634, v. 6, 319–20, MI/DS-J, CNSAS.
71 Tofan, *Antologia Radio Europa Liberă*, 277.
72 Noel Bernard to Sig Mickelson, memo, March 11, 1977, RFE/RL Romanian Broadcasting Department records, miscellany 1977–79, HILA.
73 "Romania Earthquake," Noel Bernard to Sig Mickelson, memo, n.d., RFE/RL corporate records, box 329, folder 15, HILA.
74 Noel Bernard, interviewed by Sig Mickelson, June 25, 1981, Sig Mickelson papers, audiotape 5, HILA.
75 "Romania Earthquake," Noel Bernard to Sig Mickelson, memo, n.d.

76 Noel Bernard, interviewed by Sig Mickelson, June 25, 1981.
77 Bernard.
78 James Edwards to Thomas Bodin, memo, March 9, 1977, and Nathan Kingsley to William Mahoney, memo, March 10, 1977, both in RFE/RL corporate records, box 1858, HILA.
79 Noel Bernard, interviewed by Sig Mickelson, June 25, 1981.
80 Nathan Kingsley to William Mahoney, memo, March 10, 1977.
81 "Romania Earthquake," Noel Bernard to Sig Mickelson, memo, n.d.
82 AGERPRES, "March 4, 1977."
83 Noel Bernard, interviewed by Sig Mickelson, June 25, 1981.
84 "Romania Earthquake," Sig Michelson to US senators and congressional representatives, draft letter, n.d., p. 4, RFE/RL corporate records, box 329, folder 15, HILA.
85 Johnson and Parta, *Cold War Broadcasting*, 144.
86 Noel Bernard to Liviu Floda, memo, March 15, 1977, Liviu Floda papers, box 5, HILA; "Romania Earthquake," Noel Bernard to Sig Mickelson, memo, n.d.
87 William Mahoney to Sig Mickelson, memo, March 11, 1977, RFE/RL corporate records, box 1858, HILA.
88 Mahoney; Ralph Walter to Albert Mickelson, memo, March 15, 1977, RFE/RL corporate records, box 1858, HILA.
89 Noel Bernard, interviewed by Sig Mickelson, June 25, 1981.
90 Nagat, "Ceauşescu's War," 236.
91 Tofan, *Antologia Radio Europa Liberă*, 59.
92 Tofan, 274–75.
93 C. Popa, *Elena Ceauşescu*, 57–58.
94 Popa, 276.
95 Popa, 281.
96 AGERPRES, "March 4, 1977."
97 For footage of Nicolae and Elena Ceauşescu touring Bucharest's damage, see Partidul Comunist Român, "Cutremurul din Bucureşti 4 martie 1977."
98 Secţia Cancelarie, 18/1977, 2, FCCPCR, ANR.
99 Secţia Cancelarie, 20/1977, 12.
100 Secţia Cancelarie, 18/1977, 13 rev.
101 Secţia Cancelarie, 20/1977, 17.
102 Secţia Cancelarie, 20/1977, 17.
103 ActionAid Agency for International Development, "Earthquake Sitrep."
104 Secţia Cancelarie, 23/1977, 16, FCCPCR, ANR.
105 Hewitt, *Regions of Risk*, 206.
106 They survived for forty-two hours; see *Flacăra*, March 10, 1977, 14.
107 She was trapped for 187 hours; see *Los Angeles Times*, March 14, 1977.
108 Antonios Pomonis, pers. comm., April 11, 2016.
109 Secţia Cancelarie, 18/1977, 2, FCCPCR, ANR.
110 Secţia Cancelarie, 20/1977, 17.
111 "Ceauşescu's Leadership Position, Summary," 544.
112 D.11.737, v. 105, 9, DB, CNSAS.
113 Secţia Cancelarie, 20/1977, 10, FCCPCR, ANR.
114 Secţia Cancelarie, 20/1977, 11 rev.
115 Secţia Cancelarie, 20/1977, 10.
116 Secţia Cancelarie, 20/1977, 10.
117 Secţia Cancelarie, 20/1977, 11.
118 Secţia Cancelarie, 20/1977, 13.
119 Secţia Cancelarie, 35/1977, 3; World Bank, *Report and Recommendation*, 12.
120 D.3635, v. 2, 126, MI/DS-J, CNSAS.

121 *New York Times*, March 7, 1977; *Los Angeles Times*, March 10, 1977; D.13.339, v. 37, 52, DB, CNSAS.
122 Secția Cancelarie, 35/1977, 80 rev. and 92, FCCPCR, ANR.
123 Noel Bernard, interviewed by Sig Mickelson, June 25, 1981.
124 Raab, *All Shook Up*.
125 MediaFax, "Mărturii din 4 martie.".

3 Recovering Valuables and Influencing Foreign Assistance

The Ceauşescu regime's initial disaster response focused on helping survivors and recovering victims, but a concurrent and longer focus was on the recovery, and even bolstering, of Romania's state-run economy. Because Romania was a socialist republic, the government controlled most of the economic production. Its public messages centered on earthquake survivor assistance, but privately the regime sought to repair structures important to the economy and collect fungible assets. It appropriated valuables and solicited cash and specific technical equipment. First, at the collapsed building sites, first responders searched for survivors and victims as they concurrently dug out cash, bankbooks, gold, silver, or any item the regime considered valuable, inventoried and warehoused what they found, deterred and arrested looters, returned items people could prove ownership of, and absorbed the rest. Second, while it accepted all humanitarian assistance sent, the regime communicated to foreign governments and international organizations its preference for cash, low-interest credit, and specific technical equipment over in-kind goods. Last, it ordered the Securitate to do the same, using its network abroad.

Digging through the Rubble: Valuables Found, Claimed, Appropriated, and Stolen

The regime closed the search for victims four weeks after the earthquake with the submission of its final report listing the identified and numbering the unidentified dead.[1] Six weeks after the earthquake, Minister of the Interior Theodor Coman said that the work to retrieve valuables from the rubble would continue:

> The decision was taken to continue the operation to recuperate the various valuables from the sites of the demolished buildings damaged by the earthquake. In this goal, the Interior Ministry agents must give their all in ensuring the security of these respective locations and the recovery of all valuables.[2]

DOI: 10.4324/9781032632933-4

That work continued until its official end on May 13.[3] Interior Ministry workers, specifically the Securitate and police, guarded, searched for, and inventoried anything of value they dug out of earthquake-damaged sites. Publicly the regime claimed to be searching for construction materials and culturally important items, not cash, precious metals, or savings deposit books. The state-run media reported that

> the recuperation of valuables, brick by brick, all that could help with the reconstruction work, is underway … not one small street was missed so that we do not lose anything of patrimonial or cultural-artistic value such as book and art collections, [other] rare collections, scientific documents.[4]

More than 1,500 Interior Ministry workers recovered usable construction materials, sorting bricks and collecting almost one hundred tons of scrap metals—steel, lead, cast iron, aluminum, and copper.[5] They retrieved technical equipment as well: one police regiment removed two "electronic computers" and some accompanying equipment from the Transportation and Telecommunications Ministry building valued at over four million Romanian lei.[6] Not all finds were so grand, but the Interior Ministry recorded all, including five hundred hangers and thirteen cartons of buttons from a clothing factory.[7]

Household items represented the majority of what workers dug out of the rubble. In total, they found, recovered, and inventoried more than a hundred thousand household objects as they worked at the partially and fully collapsed building sites.[8] The 1977 value of Romanians' lost and damaged possessions was estimated at $95 million.[9] The regime inventoried truckloads of items and transferred them to ten temporary deposit sites across Bucharest.[10] It also stored items in warehouses, schools, a communal bathhouse, a department store, and a grocery store.[11] Once or twice a day, the Interior Ministry updated the inventory lists and Securitate officers and army troops helped inventory the items and guard the storage sites. Nine days following the earthquake, Minister Coman established a Bucharest-level commission that comprised "the chief prosecutor, tribune chairman, chief clerk, financial administrator, and representative militia [police] bodies" to deal with the items recovered from the collapsed buildings. The city commission also had eight subordinates, one for each of the city's eight sectors. Those local commissions were charged with "continually taking measures for disinfecting, sorting, inventorying, and guarding the valuables recovered."[12]

Household items found and inventoried included gold coins, watches, necklaces, earrings, rings, wedding bands, gold and silver flatware, sterling silverware, books, stamp collections, artwork, handmade carpets, fur coats, electronics, suits, dresses, haberdashery, indoor stove natural gas tanks, refrigerators, and tools. Workers listed small electronics found, including photo and movie cameras, a film projector, radios, radio–cassette players and recorders, and telephones. They recovered religious icons, paintings, small sculptures, and porcelain vases. They also dug out more than one hundred automobiles. They found only two guns: a Browning pistol and a hunting rifle with ammunition.[13] The Securitate recorded that workers who found high-value items like "gold rings, expensive dresses, and shoes with foreign labels, furs, [and] hundreds of thousands of Romanian lei and dollars sewn inside upholstery" noted the material differences between those who lived in the city center, where most of the buildings collapsed, and "regular citizens."[14]

In the first days of the disaster response at the collapsed building sites, the Ceauşescu ruling couple lambasted the Interior Ministry for concurrently focusing on recovering items while searching for survivors' and victims' bodies. Nicolae Ceauşescu ridiculed the obsessive inventorying of items, saying,

> one broke a plate, another a beautiful vase. We cannot start counting every broken shard. Nowhere in the world do they start counting who had five broken drinking glasses, and who else lost ten. Another had a vase broken ... we cannot start including these in our calculations [of the economic losses]. We will be ridiculed. Come on, let's be serious.[15]

Elena Ceauşescu had similar complaints, saying, "each is guarding himself; they stay and count rags and shirts.... I saw many that were standing around and counting curtain rings, making an inventory. Was that the most urgent [task] to be done?" She criticized Coman, who supervised those workers:

> All the time [the workers] say that no one is giving them anything, no one is helping them. All are commanders, yet no one knows what they should do. They search for documents to identify them, but they are not concerned with the people.[16]

She recalled that in the aftermath of the 1940 earthquake, their nonsocialist predecessors had prioritized searching for people:

> During the bourgeois time the first work that was done was to get to the [victims]. We are speaking now of humanism.... The first work must be to save people and not to remove clothing and search for identity documents. You have not thought at all about the people.[17]

The World War II-era fascist government that orchestrated the 1940 earthquake recovery efforts was led by Ion Antonescu, who took power only two weeks before Romania allied with the Axis Powers and allowed Nazi troops into its territory. For Elena, the number two communist, to say that fascists were better at disaster recovery than their socialist regime leaders cut deeply.

Even with the ruling couple's complaints, the valuables recovery continued. The Interior Ministry workforce expended much effort to dig out, inventory, store, guard, and return or discard the hundred thousand household items recovered. The number of household items found points to the wealth and consumerism of Bucharesters. By the 1977 earthquake, the standard of living for most urban Romanians was comfortable. They had more disposable income, free and leisure time, and access to consumer goods than during the previous twenty years. During the early years of the Ceauşescu regime, there was a shift toward greater availability, mostly in urban areas, of consumer goods and other "scarce or unavailable luxury items and household durables from caviar and cognac to vacuum cleaners and television sets."[18] Amid the disaster recovery, the regime recovered from the collapsed buildings' rubble more than 6,000 women's formal suits and dresses, exponentially higher than the paltry 185 men's suits recovered.[19] This data points to women's consumerism. In the early decades under Ceauşescu, Romanian socialist consumerism, and specifically women's, "served as signs of socialist modernity" as the regime pointed to citizens' access to and ownership of soft durable "Western-style" goods as markers of socialist progress, success, and legitimation.[20] In the 1960s and 1970s, consumer products promoted the state's concept of the socialist ideal: "In contrast to the West, where youth, beauty, and sex were mobilized to sell products, in socialist Romania they were strategies for selling socialism." Before the earthquake, Romanian women had access to Western and Western-style goods like French perfume and Italian blouses, as well as images of Western styles in domestic magazines and foreign films.[21]

The Ceauşescu couple's initial criticism of the Interior Ministry's focus on valuable asset recovery did not deter it. While Elena complained, Nicolae switched his position and backed this practice on the fifth day after the earthquake. He told the Political Executive Committee (PEC),

> Of course, comrades, we must recover all the materials, all that is useful and able to be recovered.... In general, comrades, at all the demolished sites, everything is to be removed, move it somewhere, store it, and go and sort out all that is useful.[22]

Interior Ministry workers did just that. After ending the search for survivors and victims, they moved the rubble to the Glina trash dump in the city's southwest corner, now inside the village border of Popeşti-Leordeni. Interior Ministry workers continued searching for items in the rubble at the dump site through May 1977, focusing on what was most valuable to the regime: cash and precious metals.

With the regime's greenlight to continue digging for high-value items from the rubble, workers retrieved cash, bank savings, and bond deposit books, and more than 7,100 items made from gold and silver, including coins, jewelry, and sterling silverware and service pieces. Many of the gold items were identifiable, such as wedding rings and other jewelry found on the dead. Workers found cash, mostly in Romanian currency, tucked in the wallets and pockets of the dead and wounded and stashed inside household items like teapots and vases.[23]

They also retrieved a small amount of foreign cash, including US dollars, West German deutsche marks, French francs, Yugoslavian dinars, and Hungarian forints, totaling the equivalent of about $11,000.[24] The regime appropriated all unclaimed cash and gold coins, a small proportion of the gold household objects, and a large percentage of the silver.[25]

The overwhelming majority of monetary assets recovered from collapsed buildings was in the form of savings bankbooks held at and bonds issued by the state-owned Savings and Loans House (Casa de Economii şi Consemnaţiuni, CEC). Workers found 1,225 identifiable savings bankbooks, 62 unidentifiable savings books, and 5,496 numbered savings bonds. Their combined total value, equivalent to about $460,000, represented more than half the value of all that the regime found and inventoried from the collapsed building sites' rubble.[26] Only half of the bankbooks with identifying information were returned or claimed, resulting in the regime absorbing about $221,000 from Romanian citizens' savings accounts.[27]

It was difficult for survivors and families to claim household items, cash, and bank savings books dug out from the rubble of the collapsed building sites. First, it was hard to physically get to where the regime warehoused these materials. The destruction in the city center rerouted busses and halted trams, making transportation difficult. Distribution of the items across temporary warehouses, banks, and police stations made knowing where to go almost impossible, so people traveled to many or all locations to find and retrieve things. One month after the earthquake, the government announced the claims procedure in the newspaper, setting three months for claims from the ten warehouse sites, the National Bank, the CEC bank branches, or the police.[28] Of course, people had also tried to claim things in the month before, from items strewn on sidewalks or piled for transfer.

In Bucharest, city leaders directed the claims and distribution process. The Interior Ministry initially stored the most valuable items—unclaimed savings bankbooks and bonds, foreign currency, and gold and silver items—at its offices, before sending them to the claim sites.[29] The regime required proof of ownership, and many items were not returned either because people did not have sufficient documentation or because the regime kept those it considered valuable to the "national-cultural patrimony," including books of "historic interest," art, decorative objects, religious icons, stamp collections, paintings, statues, and "old" documents.[30] To claim a savings bond, one had to present proof of the bond number or purchase.[31] Victims' relatives, too, claimed items when identifying a deceased loved one. The regime noted that most families refused to take any clothing and less than a third took anything at all, usually "sentimental family items, books, or objects found on the dead."[32] Survivors and victims' heirs claimed almost half of the valuables recovered. The regime did not keep the less valuable unclaimed household objects found; instead, these were sent to schools, orphanages, nursing homes, the housing displaced, or "those in special situations," or simply discarded.[33] While it did return items to those who could prove ownership or inheritance rights, the regime kept the majority of the unclaimed valuable items, documenting the cash, bank deposits, gold, and silver it kept, a portion of which it deposited into the domestic "1977 Account," one of two accounts created for earthquake assistance (its sibling, the "1000 Account," was for foreign receipts). No documentation of either account has been found.

At the same time that the Interior Ministry deployed its workforce to sift through the rubble for valuables, it prevented citizens from doing the same. Looting is not uncommon in the wake of disaster, and in 1977 the Securitate took measures to prevent it. Dispatched officers and cadets kept people away from the collapsed buildings, rubble dump sites, and warehouses to protect them from injury and items from theft. They also took proactive measures to deter looting. During the state of emergency, the regime prohibited people with criminal records from traveling to and around the capital, targeting those "predisposed to commit crimes" (coded language for the racial profiling of Roma).[34] While not an indication that the crime prevention strategy worked, the police and Securitate in Bucharest arrested only forty-four civilians for looting. These individuals were caught stealing while inside the evacuated buildings and picking through the rubble of collapsed buildings and dump sites. Someone tried to steal binoculars, another a bicycle, and a truck driver tried to steal books recovered from a ruined apartment building. Police caught two men in a taxi transporting a kitchen stove, and a man who filled his car with rubble "in which it could be seen ten, fifty, and hundred Romanian lei banknotes flying around."[35] Citizens reported looters, too. A retired engineer reported that a group of "particular" individuals removed seven buckets of jewelry and other valuables from the rubble of a damaged building. Two neighbors caught someone trying to rob the apartment next door.[36]

While citizens were prohibited from approaching the damaged and collapsed building rubble piles and dump sites, regime workers had full access to steal from them while digging in or transporting the rubble. Most workers who were caught stealing had small items, slipping them into their pockets or hiding them to retrieve later. They were caught trying to steal cash, jewelry, cigarette lighters, wristwatches, pocket watches, all sorts of pens (metal, plastic, and fountain), gold and silver flatware (tablespoons were particularly popular), tubes of toothpaste, a calculator, electric razors, a thermometer, and foreign cigarettes. Workers also tried to pilfer things too large for their pockets: a hair dryer and a Czechoslovak Tesla cassette recorder. One was caught trying to steal the diploma of one of the wives of the actor Toma Caragiu, who died in the earthquake. Troops transporting in-kind humanitarian assistance were caught stealing large items like portable heaters and even cars. Three military cadets found a Romanian Dacia 1300 automobile in a pile of rubble, moved it to the front of their barracks, stripped it for parts, and dumped its chassis in a nearby pit. Their project was obvious to all in the barracks, and a fellow cadet reported them to their unit's supervisor.[37]

Like the military cadets who turned in their peers for stripping the car, Interior Ministry underlings and superiors reported looting among their ranks. How many of their supervisors looked away or demanded their cut is difficult to know, but examples of such power abuses exist. While there are a few examples of workers turning in their supervisors, the majority of the cases reported were workers concerned that supervisors were collaborating with subordinates to steal. For example, three Interior Ministry truck drivers chauffeured as their superiors took things from the rubble. In another case, a tank brigade driver stole "four RHR brand West German electric heaters and other valuables" while driving a foreign aid shipment in his tank from Bucharest's international airport to a hospital. The police searched his home and found three of the heaters, with the fourth found in the apartment of his unit's superior.[38]

When people were caught stealing, the Ceauşescu regime punished regime workers and citizens differently. Under the 1969 penal code, looting was considered theft or robbery depending on the circumstances under which it happened.[39] The code defined theft as taking a "movable" item that did not belong to oneself and allowed for a prison sentence of between three months and two years. It defined robbery as stealing with an additional circumstance—while in a group of two or more, at night, with a gun, under the influence of a "narcotic substance," with the use of lock picking or keys, or at the time of a "calamity"—and permitted a sentence of anywhere from one to five years.[40] The state prosecuted and sentenced civilian looters more quickly and harshly than it did the regime's workers, hearing civilian cases just days after the arrest. In contrast, Interior Ministry workers' cases were given an internal hearing and received lighter punishment if any at all. These cases were also often handled under the 1973 amendment to the 1969 penal code, which expanded punishment options, adding "reeducation" or "correctional labor" for noncapital crimes.[41] Minister Coman commented on the case of four Interior Ministry workers caught stealing from rubble dump sites, saying that they were given "exaggerated sentences." He advocated for the regime to "not make petty cases into serious problems" and commanded that the four be set free until their cases could be reviewed.[42] Coman then decided that the punishment for such offenses should be fixed to one year of correctional labor. He changed the procedure for accused Interior Ministry officers and subofficers, taking their ministry-issued firearms and relieving them from duty not as punishment, but rather so that they had time to organize their defense.[43] A much harsher punishment awaited the Romanian civilians who were caught for taking something from the rubble. These individuals—primarily unemployed young men from Bucharest—were prosecuted for robbery during a calamity. They received sentences of years in prison for stealing common household items: five years for stealing flatware and a magnifying glass; four years for a transistor radio and a manicure kit; three years for a radio; two years and six months for a typewriter; and one year and ten months for a pair of pants, a manicure kit, and a wallet.[44]

Cash, Please: Foreign Assistance Preferences

Giving in all its forms is common during disaster response. Foreign individuals, governments, and organizations helped Romania in the wake of the 1977 earthquake. Yet, the official amount of humanitarian assistance Romania accepted paled in comparison to the losses it sustained. The regime reported receiving cash assistance totaling $35 million and estimated the earthquake damage at $2 billion, half of which was in the housing sector.[45] The direct aid reported represented barely 2 percent of the estimated losses (and did not include the credit and low-interest loans Romania received, such as $130.3 million credit from the International Monetary Fund). Romania accepted all items sent along official channels, but it privately tried to guide receipts. Already having had significant experience receiving in-kind goods from abroad following earlier floods, the Ceauşescu regime preferred earthquake recovery assistance in the form of donated cash, credit, low-interest loans, and telecommunications and construction equipment. Publicly, the regime praised the receipt of donated "oranges, chocolate, and powdered milk for the wounded in Bucharest and perishable food items (citrus fruits, baked goods, cheese, juice, etc.) sent to ... daycare centers, kindergartens, dormitories, public cafeterias, and hospitals."[46] But, privately, the regime tried to control the in-kind assistance sent. It seized control of the assistance receipts for two reasons: first, to deter in-kind items from moving to the illegal, secondary market; and second, to bolster its economy with cash and its operations with specific technical equipment. The regime's prioritization of the economy was not completely unnoticed. US diplomats observed during Ceauşescu's one press conference with foreign journalists that he was "avoiding answering a question on the need to divert economic resources from heavy industry toward consumer goods to meet the needs of those who lost their possessions."[47]

The bulk of foreign in-kind humanitarian aid came in the form of medicine, food, bottled water, medical and technical equipment, and construction materials. UNICEF donated $25,000 worth of food, medicine, and other assistance; the American Jewish Joint Distribution Committee helped Bucharest's Jewish residents and sent relief workers; Switzerland sent an entire school; and Great Britain donated a hundred breeding sows and ten boars.[48] While the majority of in-kind assistance came from nonsocialist governments and international nongovernmental organizations, the Soviet Union sent the largest in-kind donation: the materials to build an entire factory for manufacturing prefabricated housing materials, valued at 10 million Soviet rubles or $13 million.[49] Unsolicited support came, too, from the Romanian diaspora and others with ties to Romania. The regime reported: "Some foreign citizens, many of them originally Romanian, offered money, medicine, and clothing, and expressed their solidarity with the Romanian people."[50] Experts arrived to provide technical assistance for search and rescue, as well as seismic assessment. China, Yugoslavia, West Germany, the United States, Japan, Turkey, and Greece, along with the USSR, sent seismic engineers or other specialists.[51] The United

Nations provided rescue workers.[52] The Swiss sent teams with rescue dogs who helped find buried survivors and victims.

Before the 1977 earthquake, receiving items from abroad was common, with this "economy of everyday assistance across long distances" well established. In part because of Ceauşescu's atypical open relationship with the West, non-socialist governments, nongovernmental organizations, and individuals, particularly among the diaspora, regularly sent consumer goods, food, and medicine.[53] As Luminiţa Gatejel explains, Romania's portion of the Iron Curtain was in reality quite permeable.[54] Starting after the 1970 and 1975 floods, people mailed care packages to Romania or traveled there to deliver them in person. In addition to everyday essentials like clothing and food, people sent luxury items like coffee and chocolate. Few mailed contraband, like condoms and birth control pills, which were practically impossible to find because of the 1966 criminalization of contraception and abortion; in many of those cases, they risked in-person delivery.[55] Even with illegal items, this influx of goods was so common that the regime did not restrict it. As Cristian Capotescu explains, "since the 1970s, socialist states began to tacitly accept the growing import of foreign consumer goods and currency through small border shopping or trader tourism as a means of rebalancing trade deficits with the West."[56] For relatives living in the diaspora, this ability to give was comforting and created a sense of solidarity.[57] These small one-way consumer goods networks became an important part of procuring everyday necessities and stocking the secondary economy. When disaster struck, it was natural for those same networks to activate and send humanitarian assistance.

The influx of in-kind items arriving from abroad in the wake of the earthquake overwhelmed the regime. Days after its arrival, much of it sat unused: technical and material supplies, medicine and medical equipment, and food stored in three temporary Bucharest warehouses went undistributed as officials were waiting for approval to do so.[58] Some assistance that did move was because workers stole it. Within days of the earthquake, Ceauşescu clamped down on these thefts from within the regime's ranks. It was clear to him and others in the government that many officials were mishandling, even stealing, the in-kind foreign assistance receipts under their charge. Ceauşescu addressed PEC members:

> I have been informed that there is a lot of disorganization at home ... the airport [Otopeni, where almost all the foreign assistance arrived] is like a carnival [*bâlci*] when the aid arrives, everyone wants to take it. One minister said that all that arrives is his, another says that it belongs to him. It is completely disorganized.

Workers and managers stole medicine, food, tents, and clothing for their use or for sale on the secondary market.[59] Ceauşescu worried that there was no "secured, permanent competence" at Bucharest's international airport, and that "unauthorized people were using personal relationships, trying to influence foreign representatives to give them the aid."[60] In response, he laid out new foreign assistance receipt procedures:

> I have thought that we will organize a national committee for coordinating the victims' assistance, gathering and coordinating the domestic assistance, and coordinating the use of foreign aid in a unified way. From all that arrives from abroad, none of us should release or take anything.[61]

Even with Ceaușescu's early clampdown, some foreign assistance made its way to the secondary market. During the weeks that followed the earthquake, medicine, first aid products, food, and other items intended for survivors and the displaced were seen for sale on the secondary market. The Ceaușescu regime was concerned not so much with individual thieves, but rather with networks taking and selling larger quantities of in-kind foreign assistance. This level of theft was not new for the regime. Following the floods in 1970 and 1975, midlevel managers stole and sold foreign assistance, though not on the secondary market, but rather in state-owned stores and other state-run entities, like vacation resorts. In testimony presented before the US Congress about whether to provide Romania with additional humanitarian assistance following the 1977 earthquake, several letters from Romanian immigrants living in the United States explained that following the 1970 flood, "emergency help given to them consisted only in used clothes collected from other Romanians, while goods received from outside … were sold in the village's state[-owned] stores." Emergency tents received from the United States "were seen installed on the beaches and rented to the people during the summer." And the "fish received from China, powdered milk and cocoa from Holland, Switzerland, Denmark, canned meat products, coffee, as well as materials and clothes received from other Western countries to be distributed to needy people, had been sold in state stores."[62] Following the 1977 earthquake, one reason the Ceaușescu regime insisted on cash over in-kind assistance was to limit such entrepreneurial activities.

The day after the regime established the new foreign assistance receipt and distribution procedures, it immediately suspended all requests, until it could, as it informed the US ambassador, "work out its priority needs"; it explained that the "suspension was necessary because large amounts of aid and foreign experts were arriving in Bucharest without coordination."[63] Ceaușescu created the National Assistance Committee and appointed Elena Ceaușescu as its head.[64] This added one more role to her responsibilities portfolio. In 1974 Ceaușescu had included her in the PEC, and then, three years later (just before the earthquake), he had named her a permanent member.[65] Appointing her to lead the National Assistance Committee demonstrated his reliance on her for the most trusted work, as this committee oversaw the receipt, organization, and storage of in-kind foreign assistance and also tracked domestic and foreign cash donations.[66] This appointment signaled Elena's importance among the top leadership of the regime, but more so Nicolae's mistrust of these officials.

Ceauşescu aimed to fill the two special bank accounts—the "1977 Account" for domestic donations and the "1000 Account" for foreign receipts—with cash received in the name of earthquake relief.[67] Just four days after the earthquake, he told the PEC,

> we can't do a lot with $5,000 or $25,000. If [foreigners] want to help us they can give us $50 million in credit without interest, but with $25,000 we won't get rich. Of course, we will not refuse if they contribute.[68]

The regime also did not refuse in-kind foreign assistance, but instead made specific requests. Ministries prepared lists of the medicine, medical equipment, machinery, and parts they needed and desired, sharing these lists with foreign embassies. They asked for things not only for immediate disaster recovery but for general economic advancement. As Gatejel explains,

> simply put, Romanian officials considered emergency relief as part of a larger concept of aid that also included long-term recovery and reconstruction measures. They explicitly mentioned loans and trade facilitation that went far beyond the humanitarian help directed to those affected by the [1977 earthquake] disaster.[69]

During the first few weeks after the earthquake, Ceauşescu vacillated between denying the need for foreign humanitarian assistance and making specific requests. The second day after the earthquake he declared that Romania would not make specific requests, telling the PEC,

> we will receive those people from [West] Germany and Switzerland who want to arrive with dogs and put them to work. Therefore, we will proceed in this way … we do not approach anyone to ask them for something. Not even the socialist countries. If they don't feel like it, they don't feel like it.[70]

But, Ceauşescu was also clear about what he wanted from foreigners when, practically in the same breath, he said,

> we are interested in beds, medicine, and, in the first place, food and equipment, and materials for repair. I am not referring to all but only to the developed countries that say they want to help us.… We will call the US Embassy today and tell them that we are asking for help.[71]

By the next week, however, his preference for what types of foreign aid Romania wanted became very specific.

Just two days after the earthquake, to support the regime's preference for cash and capitalize on the influx of foreign assistance, the Securitate told its network abroad: "You will identify new possibilities for assistance, use the influential relationships you have for this goal, and report these immediately to the home office."[72] Ceauşescu knew exactly what he wanted from abroad—cash and favorable loans—and he ordered the Securitate to begin the "Solidarity" operation, which used its network abroad to clandestinely encourage and solicit the same.[73] Publicly, Ceauşescu also requested cash, no-interest loans, and specific equipment during his one foreign press conference of the disaster response: he appealed for "foreign credits on favorable terms" and shared his vision for the disaster recovery, saying, "not only is the five-year [economic] plan certainly not going to be affected, but living standards will increase even more. This is our challenge."[74]

Just two days after the earthquake, an officer code-named "the Storm" (*furtună*), stationed at the Securitate's Bucharest-based home office, communicated via telegram the Solidarity operation's directives to the Securitate's network abroad: "According to the Supreme Commander's indications, the 'University' has the mission to initiate influential actions for obtaining assistance in the form of foreign currency, medicine, medical instruments, and equipment for factories and industry affected by the earthquake."[75] Ceauşescu placed the operation fully in the hands of the "University" (*Universitate*), code name for the Securitate network outside Romania.[76] In that code, a "lecturer" (*lectoratul*) headed each foreign city mission and, in turn, supervised "researchers" (*cercetătorii*), the agents and their informants. At the start of the operation, the Securitate instructed its network abroad that "there is an absolute interest in the realm of obtaining loans without interest, terms or conditions."[77] The Solidarity operation went beyond loans to solicitation of direct foreign cash and in-kind technical assistance from foreign governments, businesses, and individuals with ties to Romania. For seventeen months, from March 1977 through August 1978, the Solidarity operation engaged Securitate officers, agents, and those in their networks in as many as thirty cities in Europe, North America, South America, Asia, and the Middle East.[78] It asked officers and agents to engage

> the full potential of the operative-informative network, which will be mobilized to conduct major influencing actions for … determining networks within governments, international organizations, financial groups, political parties, important firms, banks, people of political, economic, and financial life, and elsewhere and take concrete actions to orient their thinking about the necessity of machinery, industrial equipment, and other equipment they can send to the homeland.[79]

The Securitate advised that "if government representatives, international organizations or their representatives, or important businesses … offer to assist, indicate that Romania is interested in help in the form of industrial materials and equipment and telecommunications equipment."[80] The regime discouraged Solidarity operatives from accepting in-kind humanitarian aid and food assistance, reminding them, "do not accept beds, shoes, bedding, tents, powdered milk, etc."[81]

The Securitate home office insisted that operatives not go outside their established networks and specifically not "take actions to influence companies and important people." They were to act covertly, not "advertise" their requests for aid, and be realistic: "not ask for everything under the sun."[82] They were to guide cash donors to deposit directly into the "1000 Account," and not receive the cash as was the usual procedure. The Securitate ordered its network: "Do not—repeat do not—undertake actions to collect or spend [cash], nor should networks do the same."[83] During the first week, some operatives did receive cash donations, prompting the home office to instruct them once again: "We remind you of the instructions ... concerning steering the donations in foreign cash, which will require our researchers not to take the cash, but to guide the donors to deposit the amounts in the '1000 Account.'"[84]

The regime intended for in-kind assistance solicited by the Securitate and its network abroad to support the regime itself and not necessarily earthquake victims. Requests included not consumer goods or residential repair items for survivors, but rather equipment for the regime's use. The Securitate home office shared with its network that any in-kind construction equipment donated would be targeted for a large-scale building project as indicated by Ceauşescu:

> Strictly as information, there will be a decision to form a military construction unit (around 6,000 workers) in the homeland that will work to repair the buildings damaged by the earthquake and build several thousand new apartments. In this goal, the Comrade Principal Supervisor [i.e., Ceauşescu] told us, the staff of the Solidarity operation, that we are to work to obtain as many pieces of equipment necessary for organizing this large construction group, specifically (tall) construction cranes, bulldozers, excavators, concrete preparation plants, cement trucks, cement materials.[85]

Nine days after the Securitate shared this information about the intention to start a large-scale building project, Ceauşescu resurrected an interwar plan to build an enormous Civic Center Project (Appendix A), suggesting that his inspiration for it may have come soon after the March 4 earthquake.

In addition to construction equipment, the Securitate also encouraged its clandestine foreign network to solicit donations of telecommunications equipment not produced or readily available in Romania and admitted that these were intended for the Securitate's exclusive use. It stressed that cash and favorable loans, too, if secured, were exclusively to "support only (repeat only) our line."[86] The regime circulated lists of transmission and reception equipment that included portable generators and radios, walkie-talkies, maintenance and repair meters, transmitters and receivers, car phones, radio transmitter antennae, rechargeable batteries, battery chargers, wireless equipment tool kits, and telephone cables. The paucity of such equipment in Romania at the time had hampered first responders' effectiveness during the initial days of the earthquake disaster response. The department that handled the Securitate's listening and communications equipment, the Command for Operative and Transmissions Equipment, drafted these lists and circulated them to its "University" network abroad, instructing operatives to communicate immediately if their foreign city post heads could "realize these requests."[87] The Securitate home office further indicated that though these items were specifically for its own use, it was okay to let donors think they were for the Romanian public: "Because these materials are also necessary for the *University*, you are to make every effort to obtain these special items using the donations from the special firms offering aid to the victims."[88] The extant Solidarity operation evidence does not include an accounting of the in-kind or cash assistance it secured. Extant documents suggest that the operation did not realize significant amounts of foreign cash.

In the aftermath of the 1977 earthquake, the Romanian government gave opaque information to its citizens about the foreign humanitarian assistance it received. A month after the earthquake, a public communique in the state-run newspapers obliquely mentioned the in-kind and cash foreign donations without their total value, saying: "Additionally, offers of help were received in the form of equipment and materials from some countries, institutes, businesses, and individual citizens, and assistance from the Red Cross organizations, United Nations organizations, and other international nongovernmental organizations."[89] While the amounts may never be known, the extent to which the regime organized and deployed resources toward the asset retrieval and appropriation and foreign assistance solicitation activities demonstrates its priority on exploiting the disaster response to restore and even bolster the state-run economy. These actions were in line with the Ceauşescu regime's priorities, where the state's economic health and welfare superseded that of the individuals within it.

Notes

1 Secţia Administrativ-Politică, 12/1977, 2–53 rev., FCCPCR, ANR.
2 D.11.487, v. 2, 4, DB, CNSAS.
3 Secţia Administrativ-Politică, 2/1977, 49–49 rev., FCCPCR, ANR.
4 *Informaţia Bucureştiului*, March 12, 1977, 3.
5 D.11.737, v. 105, 180, DB, CNSAS.
6 D.12.639, v. 15, 288, DB, CNSAS.
7 D.12.639, v. 15, 356 rev., DB, CNSAS.
8 Secţia Cancelarie, 43/1977, 237, FCCPCR, ANR; Secţia Administrativ-Politică, 2/1977, 49, FCCPCR, ANR.
9 World Bank, *Report and Recommendation*, 12.
10 Secţia Cancelarie, 43/1977, 235, FCCPCR, ANR; D.11.737, v. 105, 114 rev., DB, CNSAS.
11 D.11.737, v. 105, 43, 52, 74 rev., and 161, DB, CNSAS.
12 D.11.737, v. 105, 114 rev.

13 D.11.737, v. 105, 189; D.12.639, v. 15, 352–52 rev., 358, and 362, DB, CNSAS.
14 D.12.639, v. 15, 205–5 rev., DB, CNSAS.
15 Secţia Cancelarie, 35/1977, 42, FCCPCR, ANR.
16 Secţia Cancelarie, 35/1977, 10–10 rev.
17 Secţia Cancelarie, 35/1977, 11 rev.
18 Massino, "From Black Caviar," 231.
19 Secţia Cancelarie, 43/1977, 237, FCCPCR, ANR.
20 Massino, "From Black Caviar," 230.
21 Massino, 235, 237.
22 Secţia Cancelarie, 27/1977, 51 rev., FCCPCR, ANR.
23 Secţia Cancelarie, 43/1977, 235 rev.; D.13.339, v. 37, 356 rev., DB, CNSAS.
24 The Ministry of the Interior used the July 1973-February 1978 conversion rate for traded goods, which was fixed at 20 Romanian Lei per 1 US dollar, the World Bank, *Report and Recommendation*, 4; see D.11.737, v. 105, 189, DB, CNSAS.
25 D.11.737, v. 105, 114 rev. and 189 rev., DB, CNSAS.
26 D.11.737, v. 105, 189; D.12.639, v. 15, 352–52 rev. and 358 rev., DB, CNSAS; Secţia Cancelarie, 43/1977, 237, FCCPCR, ANR.
27 D.11.737, v. 105, 75, DB, CNSAS; Secţia Administrativ-Politică, 2/1977, 49, FCCPCR, ANR.
28 Secţia Cancelarie, 43/1977, 236, FCCPCR, ANR.
29 D.11.737, v. 105, 74 rev. and 189 rev., DB, CNSAS.
30 D.11.737, v. 105, 189 rev.; Secţia Administrativ-Politică, 2/1977, 36 rev., FCCPCR, ANR.
31 Secţia Cancelarie, 43/1977, 235 rev., FCCPCR, ANR.
32 Secţia Administrativ-Politică, 2/1977, 36 rev.
33 Secţia Administrativ-Politică, 2/1977, 49.
34 D.11.487, v. 10, 37, DB, CNSAS.
35 D.12.639, v. 15, 354 and 356 rev., DB, CNSAS.
36 D.12.639, v. 15, 167, DB, CNSAS.
37 D.12.639, v. 15, 298–98 rev., 303, 304 rev., and 305 rev., DB, CNSAS.
38 D.12.639, v. 15, 288 rev. and 298–98 rev.
39 Law 31/1968 kept capital punishment on the books, where it had been since 1948. It was reserved for capital crimes, and outlawed after Nicolae and Elena Ceauşescu's executions in 1989. Legea pentru punerea în aplicare a Codului de procedura penală al Republicii Socialiste România din 12 noiembrie 1968, law no. 31, November 12, 1968.
40 Codul penal din 21 iulie 1968, title III, article 208.
41 Legea pentru modificarea Codului penal al Republicii Socialiste România, law no. 6, April 6, 1973.
42 D.11.487, v. 2, 3 rev., DB, CNSAS.
43 D.11.487, v. 2, 2 and 3 rev.
44 D.11.737, v. 105, 105–5 rev. and 110, DB, CNSAS.
45 US Congress, "Romanian Earthquake Relief," 47, 66.
46 Secţia Administrativ-Politică, 48/1977, 101, FCCPCR, ANR.
47 Samuel Lyon to Ralph Walter, memo, March 15, 1977, RFE/RL Romanian Broadcasting Department records, miscellany 1977–79, box 1858, HILA.
48 D.31.160, vol. 1, 37, IFOVS, CNSAS.
49 Secţia Cancelarie, 48/1977, 98 rev., FCCPCR, ANR; US Congress, "Romanian Earthquake Relief," 43.
50 Secţia Cancelarie, 48/1977, 100, FCCPCR, ANR.
51 Secţia Cancelarie, 48/1977, 31 and 99–100.
52 World Bank, *Report and Recommendation*, 15.
53 Capotescu, "Continuing Politics," 107.

54 Gatejel, "Bargaining for Humanitarian Aid."
55 Capotescu, "Continuing Politics," 118–19.
56 Capotescu, 109.
57 Gatejel, "Bargaining for Humanitarian Aid."
58 D.31.160, v. 1, 6 rev., IFOVS, CNSAS.
59 Secţia Cancelarie, 28/1977, 11–11 rev., FCCPCR, ANR.
60 D.31.160, v. 1, 6–6 rev., IFOVS, CNSAS.
61 Secţia Cancelarie, 28/1977, 11–11 rev., FCCPCR, ANR.
62 US Congress, "Romanian Earthquake Relief," 63–64.
63 D.31.160, vol. 1, 7 rev., IFOVS, CNSAS; James Edwards to Ralph Walter, Nathan Kingsley, and Thomas Bodin, memo, March 9, 1977, RFE/RL Romanian Broadcasting Department records, miscellany, box 1858, HILA.
64 Secţia Administrativ-Politică, 9/1977, 2–4 rev., FCCPCR, ANR.
65 Dobre, *Membrii C.C. al P.C.R.*, 138–39.
66 Secţia Administrativ-Politică, 9/1977, 2–4 rev., FCCPCR, ANR.
67 Secţia Cancelarie, 43/1977, 236, FCCPCR, ANR; D.31.160, v. 1, 3 and 8, IFOVS, CNSAS.
68 Secţia Cancelarie, 27/1977, 59 rev., FCCPCR, ANR.
69 Gatejel, "Bargaining for Humanitarian Aid," 48.
70 Secţia Cancelarie, 20/1977, 16, FCCPCR, ANR.
71 Secţia Cancelarie, 20/1977, 16.
72 D.31.160, v. 1, 8, IFOVS, CNSAS.
73 For the operation's extant archive, including details on the destruction of some of its documents, see D.31.160, v. 1, 1–87.
74 The press conference took place on March 10, 1977. See the *Los Angeles Times*, March 11, 1977, A30; *New York Times*, March 12, 1977, 4; 1238/77 CN082, RFE/RL Romanian Broadcasting Department records, miscellany 1977–79, box 1858, HILA; "Situation Report: Romania," March 11, 1977, HU OSA 300-8-47:43/5-8, p. 3, RFE/RL records, OSA.
75 Telegram 4365/V, drafted March 6, 1977, at 9:00 p.m. and transmitted twice that night (at 9:45 and 11:00) to at least twenty-six cities abroad. See D.31.160, v. 1, 26, 30, and 61 (confirmation of receipt), IFOVS, CNSAS.
76 Elis Neagoe-Pleşa, email message to author, October 19, 2016; D.31.160, v. 1, 18–20, IFOVS, CNSAS.
77 D.31.160, v. 1, 20, IFOVS, CNSAS.
78 D.31.160, v. 1, 18–20.
79 D.31.160, v. 1, 24.
80 D.31.160, v. 1, 18–19.
81 D.31.160, v. 1, 3 and 8.
82 D.31.160, v. 1, 18–20.
83 D.31.160, v. 1, 8 and 20.
84 D.31.160, v. 1, 24.
85 D.31.160, v. 1, 28.
86 D.31.160, v. 1, 16 and 19–20.
87 D.31.160, v. 1, 16.
88 D.31.160, v. 1, 22–22 rev.
89 Secţia Cancelarie, 48/1977, 50, FCCPCR, ANR.

4 Mandating Benevolence

After the 1977 earthquake struck, Romanians came out and helped. They cleaned up the debris and rubble in their neighborhoods.[1] They went to local government offices and offered their assistance. At work they voluntarily collected cash and in-kind goods and worked additional shifts at factories that made items important for disaster recovery. The regime recognized these spontaneous actions. Just five days after the earthquake, the prime secretary of Braşov County described to Nicolae Ceauşescu and the Political Executive Committee (PEC) what he saw in his jurisdiction:

> There are many offers on the part of citizens to give money. Some from Steagul Roşu [a truck and vehicle factory] and Tractorul Roşu [a tractor factory] donated one week's salary. Managers of some factories offered to give one month's salary. Some factories collected things … and some bought things in the stores for children. All of this could be good if we look at how we could guide these activities to have a positive effect … and it could be very important for the population.[2]

The regime guided several different activities. Many Romanians, especially women and children, participated in weekend work brigades known as "Work Sunday." Homeowners repaired their earthquake damage. Managers added longer hours to weekday shifts and an additional one on Sunday, in addition to their workers donating cash, portions of their salary, or annual bonuses. The work brigades and self-repairs were short lived, but the additional unpaid shifts, docked pay, and cash donations continued for at least ten months, well into 1978. The regime played on the population's sense of solidarity and benevolence, building on their charitable initiatives, and used propaganda and coercion to essentially mandate participation. The 1977 earthquake's total losses were estimated at $2 billion. The Romanian people contributed significant amounts of unpaid labor and cash in the name of earthquake recovery, and the regime never gave a full accounting of it.

Just a day and a half after the earthquake, even before the regime heard how people were taking the initiative to help, Ceauşescu had already indicated that the state would ask the population to contribute to the disaster recovery, but

DOI: 10.4324/9781032632933-5

that it would do so carefully. The Sunday morning following the Friday night earthquake, Ceauşescu told the PEC:

> We will examine our financial situation and study if we have around 800 million to one billion [Romanian] lei [about $40–50 million] for assistance, including for housing, rebuilding homes, furniture, etc. Some are without clothing; some have nothing. Later we will appeal to the population for help. I don't think today or tomorrow we will.... It will be announced what we have decided: we will create an account for that help.[3]

Initially, Ceauşescu planned to solicit citizens' help by enacting presidential decrees. That did not happen. He thought all Romanians would help, telling county-level party leaders that "the Political Executive Committee decided to allocate one billion lei for helping with repairs and construction, including for those who lost everything. We will also appeal to citizens to contribute."[4]

During the first week after the earthquake, Ceauşescu announced his plan to capitalize on citizens' benevolence, privately explaining to the PEC,

> we will not oblige anyone, but at the same time, daily we must popularize the initiatives, for example, we publish about what has been gathered and donated. One is an account where individuals can deposit and another is in the factories and industrial units, where they will work and it will be recorded and organized.[5]

The policy to ask Romanians to help in the wake of a natural disaster was not new. The "good deeds" following the 1977 earthquake were similar to those that followed the devastating floods in 1970 and 1975. Just days after the earthquake, Elena Ceauşescu recalled: "It was already discussed in the city that [people] want to give a day or a week from their salary." Her husband responded, "Good, people believe that they should proceed like last time after the flood."[6]

It is not unusual for governments in the thick of disaster recovery to ask people to help, yet in the case of 1977 Romania, not everyone was fully free to decide what to do. The regime pressured people, and it was difficult to say no. That first Monday after the earthquake, the minister of forests and building materials described how some leaders at factories that made prefabricated housing materials, important for the disaster recovery, had added an extra day, saying, "we have taken measures to work Sunday to satisfy the need [for construction materials]."[7] Managers told workers to clock in for an additional shift that Sunday. This "mandated volunteerism," today known as being "voluntold," was known in communist-era Romania as "duty with pleasure" (*datoria cu plăcere*), a gesture toward the way an underling would say "with pleasure" in response to a request from a superior, recognizing that they were not asking but requiring.[8] This practice was not unique to Romania but occurred all across the communist bloc, where regimes "asked" for participation in parades and other ideologically charged manifestations, yet all knew their presence was not just requested, but obligatory. Not all forms of volunteerism under state socialism were mandated. Children and teens often volunteered as pioneers, participating in scout-like groups that centered on developing socialist citizenship ideals.[9] The performative act of "volunteering" was one way to express allegiance to the nation and commitment to socialism, or at least the appearance of it. Such mandated "charitable" actions as parade marching—or forced "volunteering" such as that undertaken by prisoners to dig Romania's Black Sea Canal (many of them dying while doing so)—tainted the understanding of the word *volunteer* in the immediate postcommunist period. As a Peace Corps volunteer in the former Soviet republic of Moldova, just six years after its independence from the USSR, I witnessed my host country language teachers respond with comical salutes when we introduced ourselves as "volunteers." They taught us how to explain in Romanian that we were in Moldova of our own free will and not forcibly sent by our government as punishment for some infraction. The Romanian workers' labor and cash donations in the name of the 1977 earthquake recovery were forms of "duty with pleasure."

Work Sunday

The first "volunteer" activity expected of Romanians occurred during those first few weeks after the earthquake, when the Ceauşescu regime asked Bucharest residents to give up their Sunday, a day off work, to clean up streets and beautify parks. These work brigades were made up primarily of women, university students, communist youth members, high school students, and school-age children. The Work Sunday activities began as early as seven in the morning and took place on at least the last three Sundays in March.[10] The first took place nine days after the earthquake, when, as the newspaper proclaimed, "in all the neighborhoods, on all the streets, the large movement of forces will expend energy to ensure effective clean-up work!"[11]

The regime continued to laud the Work Sunday efforts in state-run media. One Monday headline read, "In a Broad Effort, Youth in a Sunday Fortress of 100,000 Overalls." The article described this "broad effort":

> On what was the second Sunday after the tragic day, March 4, on the streets … there are those we have become accustomed to seeing these last nine days: busy people, in ordinary clothing, guards' uniforms, or the patriotic blue overalls, on a Work Sunday in which all the capital's residents, through self-sacrifice, worked to return life to normal.[12]

The regime played on the refrain of sacrifice for the good of the nation. After the second Work Sunday, the capital's daily newspaper boasted that about 10 percent of the capital city's population had helped, dressed "in work clothes" in "groups of tens of thousands" to clean up the city. People assisted specialists with reinforcement work at damaged building sites, children planted flowers, others helped clean up parks, communist pioneers and patriotic guards unloaded materials from trains at the railway station, and others swept up debris from courtyards and streets.[13] The Work Sunday requirements built on earlier expectations of participation in such mandated "volunteer" activities as parade marching and ushered in a new wave of "voluntary" work, from children in particular. During its disaster recovery, the regime replicated the widespread "practical agricultural" work brigades commonplace since the land collectivization in the early postwar era. These would also later be used by the Ceauşescu regime in the 1980s, when school-age children and youth made up work brigades with their school teachers organizing and supervising them as they assisted in the autumn harvest and other agricultural activities. The regime described the mandated Work Sunday efforts after the earthquake not just as beneficial but as necessary for the "return to normal, economic-social life," an essential part of the disaster response struggle. Ceauşescu was the chief commander on this "battlefront," with the Work Sunday one day in which all "citizens are employed in the battle for repairs and normalization."[14]

DIY, Construction Materials, and Payments for Housing Repairs

The regime placed another disaster recovery burden on people, asking many to make repairs to their own homes and community buildings damaged by the earthquake, excluding those that experienced structural damage. The earthquake damage to the housing sector was significant, estimated at $1 billion, affecting six hundred thousand households, or 11 percent of Romanian homes, with slightly more than two hundred thousand of those uninhabitable. Bucharest was hardest hit, with almost half a million homes damaged, or 41 percent of the capital's housing stock.[15] That June, in line with its modernization goals, the regime turned to building new housing, with a new five-year national and rural systematization plan for more than two hundred villages, towns, and cities. It included a new "large investment plan" for "over one million homes" across Romania to ameliorate the housing shortage.[16] In the year that followed, the regime faced other economic and sociopolitical challenges and left the housing problem among the last to be addressed. The housing shortages caused by the 1977 earthquake's damage remained well into 1978, when the World Bank determined that Romania needed $1 billion to repair residential damage alone, an underestimate because the regime undercounted, omitting an estimated eighty-two thousand homes damaged by the earthquake from its report to the World Bank.[17]

Five days after the earthquake, Ceauşescu told the PEC, "everyone must participate in repairing his house or apartment, of course under guidance, as every worker is able to plaster, to repair his own [home]."[18] Communities, families, and individuals were left on their own, as one newspaper reported: "Every citizen, every family, will participate in repairing the homes in which they live!"[19] Four months after the earthquake, in July, Bucharest mayor Ion Dincă mentioned that in addition to the twenty-nine thousand Bucharest workers who had left their regular work posts to repair industrial buildings, regime officials mobilized "an important number of schoolchildren and university students" to repair educational buildings.[20] Ceauşescu ordered this, saying, "the same will take place with education, grab hold of the students and their parents to rebuild the schools." He dismissed a suggestion from the education minister to select and repair two thousand schools, and instead commanded that "repairs will start at all of them. Within two months all will be repaired using the work of the local population from the cities and the surrounding towns, except a few with severe damage."[21] He ordered the same for the repair of public and cultural institutions, such as libraries, museums, theaters, and movie houses.

Ceausescu told the PEC members,

> we will give them the materials because we don't have enough [workers] to do this.... In an organized way, youth, students, and everyone will repair their apartment or house. We will give materials, prepare and show them, but they will make [the repairs]. Outside of the seismic structural reinforcement work, which must be done by specialists, the rest must be fixed by all.[22]

Even though it promised supplies, however, the regime did not give homeowners sufficient construction materials. The first Monday after the earthquake people stormed the shops, with sales for construction materials tripling, resulting in shortages. Ceauşescu and regime ministers repeatedly discussed in the days and weeks that followed the earthquake the lack of materials available.[23] They did not have enough cement panels, mesh, waterproofing sheets, windows, concrete, mineral compounds for concrete, or PVC pipes.[24] Civilians, workers on the repair teams, and even the Bucharest mayor complained that they did not have enough materials.[25] In a rare case, a state-owned construction company organized and dispatched fifty-five teams of eleven workers each to help citizens repair their homes. The teams complained, however, that they were not even able to help one-fifth of the citizens because they did not have sufficient bricks, plumbing equipment, ceramic tiles, or scaffolding. Team managers "felt that if they had the construction materials in the quantity they needed they would be able to contribute much more to repairing the damaged homes."[26] The regime then made available for purchase "construction materials for the population," which included tons of cement and plaster, two billion bricks, tens of thousands of feet of lumber, and thousands of concrete slabs, tiles, and buckets of paint.[27] Civilians bought materials to plaster and paint cracks in walls and ceilings, repair roofs, refit or replace broken pipes and wires, and mend or replace broken furniture and household items.

The regime exploited the fact that Romanian homeowners rarely outsourced home repairs. As early DIYers, Romanians made most repairs themselves. This was a reality all over the former communist bloc. The Ceauşescu regime capitalized on these skills and presented repair tasks as a civic duty. In the weeks following the earthquake, state-run newspapers published articles about the expectation that people make their own repairs, with headlines like "A Civic Responsibility Imperative: Every Family, Every Citizen, Must Participate in Repairing Their Home." One article featured a man, retired for three years, photographed fixing his roof. The paper quoted him: "We know that in such difficult situations in which hundreds of people find themselves, we cannot sit with our hands folded, waiting for others to come and repair the damage. The builders must work on rebuilding the large buildings."[28] Here the regime used state-run media and the handy pensioner as the mouthpiece for its message to preserve the engineers and workers for the larger disaster recovery tasks. The state-run media recycled this theme (and headline) about patriotic DIY repairs many times, with newspaper headlines like: "Every Citizen, Every Family to Participate in Repairing Their Home—A Burning Request, a High Civic Duty." The article accompanying this headline underscored the message, insisting that "now when every team is precious for the builders, the population's direct contribution to repair their apartments is imperative." The author described how,

> on this workday, when Bucharesters have rolled up their sleeves for fiercely continuing an impressive collective effort to remove the traces of the terrible disaster and to fight to normalize life, I met many, from many different professions, after work, at home, performing the ancient skills of masons and carpenters.

In the same article, after describing three cases of people not waiting for a repair team and expressing joy and satisfaction at making necessary repairs themselves, the author concluded, "our ancient construction vocation reveals itself in these cases, shown through the foundations of the homes rebuilt, the symbol of our never-ending persistence."[29]

In addition to contributing labor for repairs, Romanians paid for them, too. In 1977 the government had no natural disaster fund. Ceauşescu resisted extending credit or increasing assistance for residential repairs. Two weeks after the earthquake, the Council of Ministers intended to present legislation to create a fund for residential repairs in the case of future disasters. At the same time, the head of the systematization department suggested that any future disaster fund must account for seismic damage.[30] Ceauşescu responded clearly and firmly that he had allocated the equivalent of about $500,000 for such repairs, saying,

> And, I know. Do not start with me again. We established which buildings were to be reinforced and those are the ones to be reinforced. As for repairs underway now, each puts his hands to work, makes repairs, and does not wait for the state.[31]

Most people paid for the repairs from their savings, some used insurance claims, and very few paid with state-issued credit. Citizens who had home insurance policies with the state-run agency received minimal payments: a third of urban property owners and slightly more than half of rural owners who submitted claims received 1,000 lei ($50) or less; the average payment to urban holders was 2,753 lei ($138), and to rural property holders 1,294 lei ($65).[32] The government extended very little credit to pay for repairs not covered by insurance. Ceauşescu did not want to give people credit, yet some lower-level regime leaders did. Just four days after the earthquake, one PEC member, Gheorghe Oprea, deputy secretary for labor propaganda issues of the Party Industrial Committee, asked to extend credit to urban Romanians for repairs, appealing to Ceauşescu, "I ask you, please, let us take measures to ensure construction materials. Let's offer some credit for those who began to repair." Ceauşescu dismissed the idea outright: "They are to put their hands to repair without credit."[33] Ten months after the earthquake, in January 1978, Ceauşescu authorized an additional 155 million lei (equivalent of about $ 7.8 million) for repairs, with 40 percent earmarked for repairing and retrofitting public buildings and the remaining amount for repairing homes.[34]

The government also demanded that people pay for much of the retrofitting and smaller repairs to buildings damaged by the earthquake. The government reimbursed, on average, only 15–25 percent of the repair costs for home interiors, such as covering cracks in wall plaster and repairing plumbing. As for retrofitting repairs, the state-run engineering institute paid the construction companies that performed the reinforcement work 40–60 percent of the total cost, with residents responsible for the remainder. Furthermore, the regime took measures against citizens who did not pay. An engineer who worked repairing damaged buildings explained, "those [citizens] who refused to pay were put under surveillance and their salary or pension withheld."[35] Some residents feared making structural repairs on their own because they thought they might be held liable in the case of damage from a future disaster. In one case residents of more than one thousand apartments in a group of five buildings, who had begun to make their repairs, were upset and feared they might be held accountable if they were not up to code.

Mandated Overtime and Docked Pay

In addition to Work Sunday and the expected DIY home repairs, the regime extracted labor from more than seven million Romanian workers in state-run industries in the name of the 1977 earthquake recovery through mandated, unpaid overtime. Workers stayed for additional unpaid shifts for almost a year after the earthquake.[36] But first, the regime delayed implementing a planned reduction in the workweek. Before the earthquake, Romanian employees of state-owned industries worked, on average, eight hours a day, six days a week. Planned to begin in 1977, a new policy would have cut one of the six full days to half, thus reducing the regular workweek from forty-eight to forty-four hours.[37] During those first days of disaster recovery, when discussing how to involve citizens in the recovery efforts, Ceauşescu stated, "most likely, comrades, in addition to some work, we will have to think about the eventual movement of the reduced workweek [law passage] until autumn and to implement it at the beginning of next year." He punctuated this comment with the claim that "everyone will understand."[38] The motivation was to try to meet industrial production targets by late March, the end of 1977's first quarter. Publicly, the regime stated that "the consensus of the colleagues of the labor unions is that meeting and exceeding plan targets are required for the national economy."[39] The government waited until 1978 to reduce the workweek and then did so by only two instead of four hours. The two-hour reduction targeted a "small minority of enterprises, basically including only those that have a high proportion of female workers and that have fulfilled plan targets."[40] It was a reward for reaching objectives, but also a small gesture to ease the double burden held by Romanian women (and many across the world), that of full-time work and primary responsibility for domestic tasks, especially if they did not live with extended family. It was a full two years after the earthquake before the government reduced the workweek to forty-four hours.[41]

Within days of the earthquake, the regime conceived of extracting free labor from workers by having them work extra, unpaid shifts in the name of disaster recovery. On March 9, the same day this plan was discussed with the PEC, the Bucharest state-run newspaper reported that people were expected to work one longer weekday and one additional shift on Sundays to try to meet or exceed planned output levels.[42] In a phone conversation the following month, Ceauşescu explained to Leonid Brezhnev (1906–82), Soviet first secretary, that people were working longer and extra shifts because the earthquake losses were "very large." Ceauşescu boldly claimed, "we will be able to resolve this problem."[43] He described to Brezhnev the economic challenges Romania faced in the wake of the earthquake:

> In industry, we have had damage and losses in production, but the [workers] are working intensely and I believe that at the end of this month, in other words by May 1, we will be able to recuperate completely all that was lost. We must, of course, work some additional Sundays, but in these conditions, of course we must.[44]

Both men probably knew that the May 1 goal was absurd, but it was also a signal of socialist solidarity and competence, with May 1 the important Labor Day holiday. In total, the regime mandated at least fourteen unpaid Sunday shifts for earthquake cleanup.[45] As it did with the Work Sunday brigades and DIY repairs, the regime presented the extra, unpaid shifts as a civic duty and propagandized them as patriotic.

When Ceauşescu mandated that the extra, unpaid shifts continue through the end of 1977, he told regime leaders to present them as "voluntary," explaining, "we must state that until the end of the year [they are] to work ten supplementary Sundays, one Sunday each month. This will not be put in the [public] communique, I say this only for us [the regime leaders]."[46] His calculations were incorrect; in 1977 Romanians worked at least fourteen additional Sundays. Ceauşescu explained that the required supplementary Sunday shifts would not be publicly announced or presented as an order: "This initiative must not be put in an [official] communication; it will be announced separately … so as not to present it as obligatory."[47] Yet, the extra work was required, and managers interpreted the directive thus. Industry managers oversaw the plan and required people to work the additional days.

Alongside the Sunday shift, the regime also imposed extended workday shifts without additional pay in the name of earthquake recovery. As with the mandatory Sundays, the official justification was to meet production plan targets in the industrial sector. The mandated longer shifts, however, included service and other nonindustrial sectors, too. The policy had both economic objectives—righting the economy—and ideological ones. All able workers were to participate.

The additional and longer shifts were legal. In 1972, the government had amended the 1950 labor law, keeping the eight-hour workday and the six-day workweek for most workers eighteen years and older, but reducing paid vacations, from a fixed four weeks for all to a range from four to two and a half weeks. Most importantly, it allowed worksites to schedule a worker for 120–360 annual overtime hours (or between 2.5 and 7.5 hours weekly) for "special circumstances in connection with production and labor interests," such as the 1977 earthquake disaster recovery.[48] The 1972 labor law did put some constraints on what managers could ask of workers. Ministers requested exemptions to the overtime limits for people they had deployed from "urban planning institutes, county-level construction organizations, ministry executive departments, factories, institutes, and several citizens … [to work at] three severely damaged areas—Bucharest, the industrial city Zimnicea along the Bulgarian border, and the town Roşiorii de Vede."[49] Because the additional and extended shifts for earthquake recovery fell within legal labor limits, there was not much workers could do, except work the shifts and quietly or openly complain.

Within weeks people were exhausted from working the longer and additional shifts, and some complained. The Securitate took note, reporting that "it is very difficult now for people to work the twelve hours and Sundays because they are tired after working that schedule."[50] It noted that the extra hours "will lead to physical and mental exhaustion" and used decreased efficiency and production outputs as an argument against the mandate.[51] The Securitate recorded an accountant who, in discussion with other workers, said that "one's health goes down if you don't give anything back to people."[52] After three weeks of mandatory Sundays and longer hours, some tired workers just up and left their shifts at the day's regular end. "The majority of workers at the Mechanical Section I of the Heavy Machinery Factory did not work the twelve-hour [shift] … because they are tired and can no longer work this lengthened program," and three people on "one day left work after eight hours saying they had difficult situations in their families." In another case, managers even locked people inside their workplace, as the Securitate reported: "The security guard closed the gate at the orders of the management, so the workers could work overtime."[53]

Many sustained the mandated longer and extra shifts for months. Some resisted. By autumn 1977 some workers at the "23 August," a heavy machinery factory, went on strike, demanding to be paid for the two hours of weekly overtime they had put in since the earthquake struck six months earlier. The US ambassador reported to the US Department of State, "management has taken the tack that even now approximately two hours of any time worked overtime should be considered an earthquake relief contribution."[54] That September, US Secretary of State Cyrus Vance suggested some additions as they drafted the Romanian section of the 1977 NATO report:

> Some of the following ideas might also be added: "the Romanian worker is being asked to work harder and longer while the prospective rise in his living standards is slowed.... Fulfillment of the housing goal in the [1971–75] plan depends on increased investment efficiency and use of private funds. Workers' salaries and savings have been docked to pay much of the cost of repairing both industrial and housing damage wrought by the earthquake of March 4."[55]

In February 1978, the US Embassy in Bucharest reported to the US Department of State that "workers have been required to 'volunteer' one day of work per month in addition to their regular 48 hours [per week]."[56]

In addition to mandating unpaid longer weekday and Sunday shifts, the regime encouraged workers to give cash, a portion of their salaries, or part of their 1976 bonus in the name of earthquake recovery. Both workers and regime members knew that most donations were not voluntary. The Securitate reported that "to come to the aid of the [earthquake] survivors, all working people across the country, including the survivors were convinced to give one [month's] salary (docked at their workplace in ten equal monthly installments)."[57] This "voluntary" initiative brought in significantly more cash than the other domestic earthquake recovery efforts or foreign assistance. Just one month after the earthquake, on April 8, Elena Ceauşescu's National Assistance Committee reported that "the commitments from workers' collectives totaled more than 3.368 billion lei," the equivalent of $168 million. More concretely, the regime had deposited the equivalent of $20.7 million in the state bank from these efforts: slightly more than half, $10.6 million, came from workers, one-third from industrial workers' collectives, 5 percent each from individual citizens and agricultural collectives, 2 percent from women and youth organizations; and just 1 percent from cultural and sporting organizations.[58] Nicolae Ceauşescu also encouraged direct cash donations, saying that because many were donating on their own through their workplaces the regime "must begin to encourage these initiatives."[59] As he did with foreign assistance, he discouraged giving items like food, clothing, or household goods—rather, he directed the regime to use advertisements and propaganda to encourage Romanians to give cash. The government never passed a law (nor did Ceauşescu sign a decree) mandating direct, salary, or bonus "donations," but people contributed because of managerial pressure and state-run media propaganda. There was a uniformity among the reported cash donation cadences and amounts, indicating that managers orchestrated their workers' "donations."

One week after the earthquake, during the first meeting of the National Assistance Committee, Elena Ceauşescu explained the process for collecting and depositing Romanians' labor and cash contributions in the name of earthquake recovery. Two days after the earthquake, the regime had opened the "1977 Account" at the state-owned bank specifically for

> voluntary contributions from workers' salaries and 1976 bonuses to help the victims … for the donations that will be received from the homeland, those from institutes, factories, and industry. It will only be done through individual or factory accounts. It will be centralized and each will decide how much they will give from their salary.… Everything should be done in an organized manner, by the party organizations, not only because it is domestic help, but [because] it is a political action of our whole nation's solidarity in the relief assistance.[60]

Official channels described the cash, labor, salary, and bonus deposits to the domestic "1977 Account" as voluntary, with Nicolae Ceauşescu telling PEC members, "we will not recommend to [give] one day or a week of work for this account, but the people will know, and those who want, can deposit into this account."[61] The regime created a second account, the "1000 Account," for foreign contributions. Ceauşescu explained that this move was to discourage any idea that contributions to the domestic account were mandatory; he did want to "amplify it as obligatory."[62] A week after the earthquake he told foreign journalists that the "losses to industry and agriculture caused by the earthquake could be made up in harder work by laborers."[63] At first Elena Ceauşescu, too, believed that the regime should not mandate cash contributions. She described her idea for a possible approach: "to start with, [tell] everyone: 'I give ten lei, another gives one hundred.'" Her husband critically responded, "this is not a collection, each will deposit what they want."[64] Just as he vacillated on the question of digging out, inventorying, and saving household items, cash, and other fungible assets from the collapsed buildings' rubble, he first proposed allowing people to choose whether to participate in donating labor and cash, but soon after changed his mind and encouraged industry managers and others to mandate unpaid shift work, cash donations, salary deductions, and bonus contributions in the name of earthquake recovery.

To popularize these "donations," state-run newspapers published daily reports of the amounts given by factories and other state-run industries as well as instructions for how to give.[65] There was uniformity among the salary deduction and bonus contribution amounts: either one day a week or month, 2 percent of salary for the remainder of the year, or some portion or all of the 1976 bonus (which was to be paid at the end of March). The regime recommended committing to give for a long time, through the end of the year or ten or eleven months. Soon after, workers at the Griviţa Roşie plant (a train and vehicle parts producer) and at the "23 August" factory "agreed to give 2 percent of their salary for the remainder of the year."[66] The director of the Metallurgy Research Institute stated that their cooperative had "decided to donate all of their 1976 bonus." Elsewhere, more than two hundred workers "decided to contribute a day a month of work and 13 percent of their 1976 bonus … [and] the union of the Factory of Touristic Automobiles agreed to give half of their 1976 bonus."[67]

The newspapers also published donation amounts from individuals, mostly those who worked outside of industry like educators and health-care workers, although these accounts were not as frequent as those about worker collectives' donations. A Bucharest University professor agreed to give one month of his salary; workers at the Union Department Store collected and donated a total of 4,000 lei.[68] Another professor from a technical school pledged to give 25 percent of his monthly salary; a pensioner pledged one day from her monthly pension.[69] During March, the only time for which we have any accounting, donations from individuals outside of worker collectives totaled only 2 percent of the $168 million deposited. The regime highlighted children's benevolence, too: "Students from General School no. 149 decided to deposit from their savings, about 10,000 lei to help victims."[70] Two different groups of pioneers "collected 1,460 lei from heartfelt children … and the pioneer group … deposited 750 lei."[71] The Ceauşescu regime knew that Romanians were curious about what their fellow citizens were doing and would follow their lead. While workers' collectives and individuals gave, cultural and sports groups organized benefit performances. The Opera House, the National Theater, and the Philharmonic pledged benefit performances.[72] Several Bucharest city soccer clubs donated match proceeds; players and staff from the Olimpia, Rapid, Progresul, and Voinţa teams each collected donations totaling about $3,000, with some players pledging a part of their salaries.[73] In contrast, the players of the Interior Ministry's soccer team Dinamo pulled together about five times their salary contributions.[74] It was one part of the Interior Ministry's large donation, over $3 million, about which it boasted: "On this occasion expressing their profound devotion and attachment toward the Romanian national interest, the Interior Ministry personnel pledge to contribute … the sum of 61,510,000 lei."[75] These group donations, like those from individuals, were a fraction of the total collected, representing only 1 percent of the total donations deposited within the first month of the effort.[76]

The Securitate documented workers' thoughts and complaints about the "voluntary" contributions. For example, it reported that

> because of the development of the work program in three shifts, on Sundays, and the reduction in 1976 bonus for victims … [the workers] are unhappy because the factory's leadership and secretary of the Party Committee took these measures without consulting the workers in

> advance; [workers said] the factory leaders do not have the right to their bonuses, and since not all employees received this [bonus] benefit not all are contributing equally to the effort.

A worker protested, saying, "I am not giving one penny. I stayed long enough after my shift; I've given my contribution."[77] Some speculated that the regime was hoarding cash—especially from foreign assistance—and extracting more from the population than was needed to address the disaster recovery. A month after the "1977 Account" opened, the Securitate recorded that among workers at the Bucharest Construction Company, "there were some really rough discussions, with shouts and insults, stating that 'the state will profit from this, taking more than was lost from the earthquake.'"[78] A few weeks later, the Securitate recorded that a pharmacist had

> insulted and expressed himself about the "1977 Account," saying that the costs for the victims are covered by foreign assistance and it should not be necessary for them to take two days from our salary until the end of the year and 500 lei of the bonus.[79]

International Transit Group workers were unhappy about the mandated donations, too, although the state-run newspaper lauded their efforts:

> The entire truck drivers' collective, engineers, and mechanics … decided to donate a day from their monthly salary, and around one thousand truck drivers, because they work abroad, will each give $20. Therefore, the contributions of this hard-working collective means, in total, over 250,000 lei and a collection of around $20,000.[80]

The truck drivers regularly drove outside of Romania and received a per diem in foreign currency, illegal for most Romanians to hold; not all donated this money voluntarily. The Securitate noted their discontent. One driver felt that the policy was "arbitrary" and "not well thought out," and said that those who drive trucks to foreign countries "have a small per diem."[81] Another driver allegedly said, "if it is voluntary then there is no obligation to contribute" and "the measures taken by the state leaders for helping the victims are made only for the eyes of the world, in the end, however, the [Romanian] population will give all of the support."[82] Some workers were upset because they did not agree with (and were not a part of the decision-making process about) the salary and annual bonus reductions and the mandated extra shifts. An engineer at the Bucharest Telecommunications Department said:

> Along with the fact that we are obliged to work late, including on Sundays, I don't know who decided that we should contribute our bonuses and a day per month until the end of the year without consulting our opinion, without asking us.[83]

More than a month after the "1977 Account" opened, the Securitate recorded a nurse at a Bucharest hospital "complaining that the hospital union created an agitated atmosphere because 'those above decided to take 800,000 lei from those who had bonuses and a day from their salary for the victims.'"[84] At the Dacia automobile factory, the Securitate reported that a worker said,

> I do not understand what kind of democracy this can be. How can the management make claims on my money? What, didn't I also suffer from the earthquake? I need to make 3,000–4,000 lei in [home] repairs. I am not saying I will not contribute, because there was a lot of misery, but not all of my bonus because I need it.[85]

Workers stated that they, too, were victims with repairs to make and shouldered an unfair burden in comparison to their white-collar managers and peers working outside the industrial sector. The Securitate reported that one worker asked, "why should I give something to those in the center who gathered money into a sack while I do not have any food to give my children?"[86] The Securitate reported information gathered about workers at Bucharest's Urban Planning Department; when told that they needed to work an additional eight hours on Sundays and another ten to twelve hours during the week, as well as contribute 2 percent from their monthly salary and half of their 1976 bonuses, "some among those [workers] asked a colleague of theirs why they, also victims, 'are obliged to contribute financially.'"[87] Furthermore, the Securitate noted that "the introduction of a longer workday among the teaching faculty of the Exterior Commerce Department has birthed unhappiness and created a state of tension."[88] The report also mentioned that an accountant working in Bucharest industry "was revolted by the fact that salaried workers are obliged to work Sundays and also must contribute part of their monthly bonuses into the '1977 Account.'"[89] Some thought that their managers, and not the regime, had initiated the collections for the "1977 Account." A textile factory worker commented: "It is unfortunate that the secretary general of the Romanian Communist Party is filled with so much concern for the people, but he does not know the games those under him are playing."[90] The lack of a specific order, presidential decree, or other legislation with precise parameters left managers free to interpret the "voluntary" contributions as they saw fit. Reflective of the redundancy in Securitate files, a painter at a state-run automobile manufacturer made a similar comment as the textile worker's about the contributions: "It is too bad that the secretary general of the Romanian Communist Party showed so much compassion for the people, but he does not know what those under him are doing because he is misinformed."[91]

It appears that blue-collar or lower-salaried workers did contribute a larger share of their salary and bonuses than did white-collar workers or managers. Laborers felt they could not afford the contributions asked of them because they were not paid enough, in particular in contrast to their managers' salaries, benefits, and perks. The Securitate quoted a worker who said, "no one is the master of anyone else's wallet. The bosses who earn 10,000 lei per month can pay until the end of the year, but a worker is not able to."[92] Another Securitate officer noted a high school teacher who said that

> he was not required to give money to the victims because "I have a public servant's pittance salary. Those who should give are the leaders who have 10,000 lei a month, eat free at the cafeterias, and have special stores where they can buy the very best at a low price; they have free services, which are like a state within a state, and the rest are as if they are not people."[93]

The Securitate's complaint reports show that workers were well aware of the inequities between them and management and other regime higher-ups, who had higher salaries, access to better housing, and permission to shop in exclusive stores.

Industrial and cultural managers, university and school directors, workers' collective leaders, and those in managerial positions organized the pledges, collections, and deductions for the domestic earthquake relief account. They also mandated the longer and extra Sunday shifts. They pursued workers for their pledges. In a report filed the day after the official formation of the "1977 Account," the Interior Ministry's national firefighter chief, Gheorghe Briceag, lauded the efforts, saying, "under the direct oversight of the party departments and organizations, [the firefighters' department] is pursuing military enrollment lists and recording the amounts officers wish to deposit into the … account."[94] It was these middle managers and directors who helped translate Ceauşescu's "voluntary" program into mandatory donations from workers and citizens.

Conclusion

These efforts inspired the regime to codify citizens' roles in future disaster recovery. The following year, the government passed the Civic Defense Law, mandating participation from all citizens and expanding the Securitate's role in future disaster recovery efforts. It outlined specific tasks for "state ministries, other central agencies, and local bodies" and obligated "citizens, state bodies, public organizations, and socialist units," in the wake of disaster, to assist in the protection of the population and the state's property. In addition to chairmen of the Romanian Communist Party (RCP) committees and executives of the people's councils, the law named government ministers and municipal, town, and village department chiefs as defense unit commanders.[95] It bound all citizens to the state's future civil defense measures, stipulating that

> all citizens, men, and women, irrespective of their [ethnic] nationality, must prepare and know the rules and measures of civil defense; to participate, if necessary, in all actions aimed at ensuring the normal operation of economic and social activity and the protection of citizens and property to ensure the normal course of economic, political, and social life.[96]

It required citizens to acquire undefined knowledge and skills to prepare for any future civil defense actions, follow measures prescribed by the regime leadership whether in peacetime or in war, participate in recovery efforts, engage in all necessary means to secure shelter, allow for housing construction on property they occupied (state-owned or private), and procure a gas mask and antichemical kit at their own expense.[97] Enlisting and obligating all Romanians in possible future civil defense measures was a remarkable new step for the regime and points to its recognition of both its own limited capability to respond to such crises and its ability to mobilize people in state-orchestrated disaster recovery efforts.

But, during the 1977 disaster recovery efforts this appeal to the populace came primarily through their unpaid labor, salary deductions, and cash contributions. The goal was to restore the economy to pre-earthquake levels and leave housing repairs to homeowners. The regime prioritized the state-run economy by keeping resources—labor and capital—for industry. As Radio Free Europe (RFE) reported one week after the earthquake, "the whole nation has been mobilized to turn out supplementary production to compensate for losses so that the annual and five-year plans can be fulfilled."[98] The largest imposition came in the form of the mandated unpaid labor and salary deductions, which continued well into 1978. There was no final accounting of the "1977 Account" fund balance, but within a month after the earthquake, the regime reported pledges of at least $168.4 million and actual contribution deposits in the name of earthquake relief equivalent to $20.7 million from Romanian workers and citizens.[99] In 1978 the regime stated publicly that it had received $50 million in direct cash from foreign assistance, but it never disclosed the amount of cash and its equivalent in unpaid labor it collected from Romanians. On the first anniversary of the earthquake, RFE journalists reported that no account balance for the "1977 Account," existed stating,

> the people who were obliged to donate a part of their salaries to this account have the right to know its destination and the amount collected. Now, a year after the cataclysm of March 4, 1977, would be the time … to publish an account balance … it's not only a moral obligation, but also a legal one … normal procedure is to publish a balance with the results of how the funds were administered.[100]

The official regime narrative about the earthquake recovery citizen contributions was one of sacrifice and "duty with pleasure." It popularized unpaid shifts and cash donations as an important part of one's civic duty to the 1977 earthquake disaster recovery. State-run newspapers lauded those efforts with headlines like, "The Entire Romanian Population Is Working Heroically to Alleviate the Earthquake's Destruction." The Securitate, however, recorded citizens' and workers' private discontent as they tried to recover from the damages brought by the earthquake, like that of a physicist at the Atomic Physics Institute, who said, "they have forgotten the misery we all went through and on all occasions speak about recovering the losses and performing superhuman work. It is as if we are not people."[101]

Notes

1 Carol Levy, pers. comm., August 8, 2023.
2 Secţia Cancelarie, 27/1977, 46 rev., FCCPCR, ANR.
3 Secţia Cancelarie, 20/1977, 17.
4 Secţia Cancelarie, 20/1977, 17.
5 Secţia Cancelarie, 28/1977, 11 rev.
6 Secţia Cancelarie, 27/1977, 60 rev.
7 Secţia Cancelarie, 27/1977, 48–48 rev.
8 Olga Ursu Stefan, pers. comm., November 18, 2016.
9 D. Georgescu, "Small Comrades."
10 Announced in *Informaţia Bucureştiului*, March 14, 1977, 2 and 4; March 19, 1977, 1; and March 26, 1977, 1.
11 *Informaţia Bucureştiului*, March 19, 1977, 1.
12 *Informaţia Bucureştiului*, 4.
13 *Informaţia Bucureştiului*, 4.
14 *Informaţia Bucureştiului*, 1, 3.
15 World Bank, *Report and Recommendation*, 13.
16 World Bank, 51–56.
17 Georgescu and Pomonis, "The Romanian Earthquake," 2.
18 Secţia Cancelarie, 27/1977, 54 rev., FCCPCR, ANR.
19 *Informaţia Bucureştiului*, March 19, 1977, 1.
20 Secţia Economică, 78/1977, 5 rev., FCCPCR, ANR.
21 Secţia Cancelarie, 35/1977, 43, FCCPCR, ANR.
22 Secţia Cancelarie, 27/1977, 54 rev.
23 Secţia Cancelarie, 27/1977, 24 rev. and 25 rev.
24 Secţia Economică, 198/1978, 69 rev., FCCPCR, ANR.
25 Secţia Economică, 78/1977, 6; D.13.339, v. 37, 326, DB, CNSAS.
26 D.13.339, v. 37, 190, DB, CNSAS.
27 *Informaţia Bucureştiului*, March 12, 1977, 2.
28 *Informaţia Bucureştiului*, 2.
29 *Informaţia Bucureştiului*, March 15, 1977, 3; see also March 19, 1977, 4.
30 Secţia Cancelarie, 35/1977, 6, 42 rev., FCCPCR, ANR.
31 Secţia Cancelarie, 35/1977, 42 rev.
32 Secţia Cancelarie, 73/1977, 157 rev.
33 Dobre, *Membrii C.C. al P.C.R.*, 443; Secţia Cancelarie, 27/1977, 33, FCCPCR, ANR.
34 Secţia Economică, 198/1978, 4, FCCPCR, ANR.
35 Transcription of *Europa Liberă* bulletin no. 412, broadcast on October 15, 1984, with the text of the letter signed by "Romanian tourists in the Federal Republic of Germany"; D.21, v. 55, 113, DB, CNSAS.
36 Action EUR, "Shortened Work Week."
37 World Bank, *Report and Recommendation*, 15.
38 Secţia Cancelarie, 27/1977, 51 rev., FCCPCR, ANR.
39 World Bank, *Report and Recommendation*, 15.
40 Action EUR, "Shortened Work Week."
41 Georgescu and Pomonis, "The Romanian Earthquake," 12.
42 *Informaţia Bucureştiului*, March 9, 1977, 6.
43 Secţia Cancelarie, 48/1977, 137, FCCPCR, ANR.
44 Secţia Cancelarie, 48/1977, 137.
45 World Bank, *Report and Recommendation*, 14.

46 Secţia Cancelarie, 27/1977, 59, FCCPCR, ANR.
47 Secţia Cancelarie, 27/1977, 59.
48 Codul muncii din 1972, law no. 10, November 23, 1972, article 119; and see Codul muncii din 1950, law no. 3, May 30, 1950.
49 Secţia Cancelarie, 50/1977, 2, FCCPCR, ANR.
50 D.13.339, v. 37, 254 rev. and 289, DB, CNSAS.
51 D.12.639, v. 15, 181 rev., DB, CNSAS.
52 D.13.339, v. 37, 254 rev., DB, CNSAS.
53 D.13.339, v. 37, 289 rev. and 344 rev.
54 Action EUR, "Reported Work Stoppage."
55 Action EUR, "ECONADS."
56 Action EUR, "Shortened Work Week."
57 D.21, v. 55, 112, DB, CNSAS.
58 Secţia Cancelarie, 48/1977, 98, FCCPCR, ANR.
59 Secţia Cancelarie, 28/1977, 11 rev.
60 Secţia Administrativ-Politică, 9/1977, 2 rev.–3, FCCPCR, ANR.
61 Secţia Cancelarie, 27/1977, 59, FCCPCR, ANR.
62 Secţia Cancelarie, 27/1977, 59.
63 1238/77 CN082, RFE/RL Romanian Broadcasting Department records, miscellany 1977–79, box 1858, HILA.
64 Secţia Cancelarie, 27/1977, 60, FCCPCR, ANR.
65 *Informaţia Bucureştiului*, March 11, 1977, 2.
66 *Informaţia Bucureştiului*, 2 and 4.
67 *Informaţia Bucureştiului*, March 15, 1977, 2 and 4.
68 *Informaţia Bucureştiului*, March 12, 1977, 5.
69 *Informaţia Bucureştiului*, March 14, 1977, 4.
70 Secţia Cancelarie, 48/1977, 98, FCCPCR, ANR.
71 *Informaţia Bucureştiului*, March 19, 1977, 2.
72 *Informaţia Bucureştiului*, March 11, 1977, 2; and March 19, 1977, 2.
73 *Informaţia Bucureştiului*, March 15, 1977, 2.
74 *Informaţia Bucureştiului*, March 17, 1977, 2.
75 D.11.487, v. 10, 40, DB, CNSAS.
76 Secţia Cancelarie, 48/1977, 98, FCCPCR, ANR.
77 D.13.339, v. 37, 289 rev., DB, CNSAS.
78 D.13.339, v. 37, 344 rev.
79 D.13.339, v. 37, 344.
80 *Informaţia Bucureştiului*, March 14, 1977, 4.
81 *Informaţia Bucureştiului*, 4.
82 D.12.639, v. 15, 210, DB, CNSAS.
83 D.13.339, v. 37, 254 rev., DB, CNSAS.
84 D.13.339, v. 37, 344.
85 D.13.339, v. 37, 290.
86 D.13.339, v. 37, 289 rev.
87 D.13.339, v. 37, 344 rev.
88 D.12.639, v. 15, 205, DB, CNSAS.
89 D.13.339, v. 37, 289, DB, CNSAS.
90 D.13.339, v. 37, 290.
91 D.13.339, v. 37, 290.
92 D.13.339, v. 37, 344.
93 D.13.339, v. 37, 344 rev.
94 D.12.639 v. 15, 15, DB, CNSAS.
95 Lege privind apărarea civilă în Republica Socialistă România, law no. 2, March 24, 1978, chapter 4, articles 29–36.

96 Lege privind apărarea civilă, articles 2 and 10.
97 Lege privind apărarea civilă, article 14.
98 "Situation Report: Romania," March 11, 1977, HU OSA 300-8-47:43/5-8, p. 5, RFE/RL records, OSA.
99 Secţia Cancelarie, 48/1977, 98, FCCPCR, ANR.
100 RFE/RL broadcast records, box 3796, folder 2, 6–8, HILA.
101 D.13.339, v. 37, 289 rev., DB, CNSAS.

5 Assessing the Structural Damage

The spring following the 1977 earthquake, construction experts and regime officials engaged in several concurrent disaster response activities to assess the building damage, especially in Bucharest. Across Romania, the 1977 earthquake's vibrations damaged more than 660,000 homes, with more than a quarter deemed uninhabitable.[1] Of those that experienced significant structural damage, 90 percent were in Bucharest, leaving thousands of multistory residential and commercial buildings unsafe. From March through June, regime officials looked to investigations made after the devastating 1940 earthquake in Bucharest to learn why so many buildings collapsed in 1977. Construction experts assessed the damaged buildings to recommend evacuations, prioritize repairs, and draft reinforcement plans. Both groups debated the merits of demolition versus repair. Nicolae Ceauşescu first supported repair over demolition, demanded fewer evacuations, and established May 1 as the deadline for assessments to be completed (in time to celebrate the important international socialist Labor Day). Over those four months, however, Ceauşescu's perception of "chaos" and "panic" from the deployed assessment commissions, evacuations, and repairs inspired him to introduce incremental changes in the regime's disaster response for the damaged buildings. In July, he abandoned his original position and ordered the end of assessments and repairs.

Past Negligence and Future Risk Revealed in the 1940 Earthquake Reports

Almost immediately after the 1977 earthquake, Ceauşescu and the Political Executive Committee (PEC) started an investigation into why thirty-two buildings fully or partially collapsed in Bucharest. They turned to investigations made after the last earthquake disaster response, following the November 1940 major earthquake. Just days after the 1977 earthquake, Prime Minister Manea Mănescu told his boss,

> Comrade Ceauşescu, I want to inform you that I found a study of Professor Beleş—I will show it to you—he indicated all the buildings, in

DOI: 10.4324/9781032632933-6

> the case of an earthquake, that would be damaged and proposed which should be demolished: they are the same buildings that collapsed.[2]

Mănescu was referring to Aurel A. Beleş's 1941 report *Cutremurul şi construcţiile* (The Earthquake and Structures), one of several investigations about the structural damage after the earlier earthquake, which struck on November 10, 1940, at 3:39 a.m. and registered 7.7 on the Richter scale, the strongest to shake Romania—and Europe—in the twentieth century. The devastation was unprecedented; the quake killed an estimated 700 people, injured at least 1,200 more, damaged 185 structures, and caused Europe's first full collapse of a building made with reinforced concrete. One hundred and forty people died when the building, named the Carlton, collapsed.[3] A mixed-use, art deco structure, it was the tallest among dozens of multistory buildings constructed during the interwar era in Bucharest. Finished in 1938, it had a movie theater at its rear, shops on the ground floor, and apartments for more than two hundred residents in the twelve stories above.

The government in 1940, headed by Ion Antonescu (1882–1946) of the fascist Iron Guard, who led Romania's participation in the Holocaust, studied why the Carlton collapsed. Experts determined that the building's support structure was insufficient, wrongly placed, or of the wrong size and shape.[4] Its corner location contributed to its vulnerability. The government supported several investigations into the reasons why certain buildings collapsed and the structural integrity of those still standing. Beleş, a prominent Romanian structural engineer and professor, authored the seismic risk assessment of the multistory buildings in Bucharest. He identified specific buildings at risk and determined that private builders had used reinforced concrete without adequate support, as the interwar era saw a "new generation of high-rise reinforced concrete structures … which seemed to be made of a magic material, reinforced concrete, good for any purpose."[5] After World War I Romanian builders had exploited the qualities of reinforced concrete but cut corners when they erected these new eight-, ten-, and twelve-story mixed-use buildings. Beleş recommended that architects calculate resistance, builders follow plans, and developers prioritize using high-quality building materials over maximizing profits.[6]

The Antonescu government also invited two German scientists to assess the 1940 earthquake damage. Their report, too, highlighted the risks associated with high-rise buildings made with reinforced concrete.[7] In 1941, Theodore Achim, a structural engineer and worksite supervisor for the Carlton building, authored his own a report about the building's collapse. He noted,

> it must be emphasized that the group of several hundred tall buildings built in the center of Bucharest, mostly in 1930–1940 decade, had the structural system made of reinforced concrete frames infilled with masonry walls. Those buildings often have a weak ground floor (for

> shops, restaurants, cinemas, etc.) and have been designed without considering, in any way, the possibility of an earthquake.[8]

Specialists had the technical knowledge, but this knowledge had not guided construction practices. Starting as early as 1903 Romanian engineers, professors, and other specialists discussed, published, and gave university courses on building with reinforced concrete. The 1925 German standards for building with it were available to them, and experts suggested regulations for building construction and permit issuance.[9] Even so, a 1926 Romanian technical journal article claimed that builders were using reinforced concrete without considering its seismic resistance.[10]

Before the 1940 earthquake, Romanian architects, civil engineers, and other experts knew about and suggested seismic and other supporting regulations for building with reinforced concrete, but private builders did not apply that knowledge and the government did not require its application.[11] Interwar authorities did not enforce the building codes. Regulatory and permit issuance policies were slow to come, and enforcement was scattershot, with little to no government oversight. Private construction companies were free to pick the civil engineer to assess and approve their plans, a factor that may have contributed to corruption and work-arounds.[12] Between 1935 and 1940 the Romanian government concrete resistance requirement levels were at their lowest rate ever, down from 1910, and were half the rate recommended by German specialists at the time.[13]

In December 1941, informed by these post-1940-earthquake studies, the Antonescu government published a draft of its first earthquake resistance technical regulations. Political disruption from war, Nazi collaboration, a coup, and Soviet occupation may all have contributed to why these regulations were not fully implemented. During World War II Romania moved from a neutral party under a monarch, to one ruled by the fascist Iron Guard. A month before the 1940 earthquake, Antonescu allowed Nazi troops to occupy Romania, and almost two weeks later he allied with the Axis powers. In August 1944, King Carol's son, Michael, with help from Romanian communists, staged a successful coup. With King Michael installed as the figurehead, Romania switched its allegiance from the Axis powers to the Allies. The 1939 Molotov-Ribbetrop Pact put Romania in the Soviet sphere of influence, and in September 1944, Soviet troops occupied Romania. In May 1945, the post-World War II government amended and published the construction technical guide. In 1948 the government established a new earthquake engineering institute, which drafted new seismic zone maps for the country and the capital.[14] Subsequent governments began to apply earthquake resistance knowledge and follow building codes. The communist government amended post-1940 building regulations in 1952, rewrote them in 1963, and revised them again in 1970.[15] A month before the 1977 earthquake the Ceauşescu regime opened the Institute of Earth Physics and Seismology. During the communist era, new construction finally considered Romania's high-risk seismic conditions.

The interwar government's limited regulations and lax enforcement of what did exist for new construction contributed to the number of collapsed and damaged buildings in the 1940 and 1977 earthquakes. Scientists have since identified a combination of factors that contributed to the full and partial collapse of the buildings in Bucharest in 1977. They mimic some of the findings about the 1940 earthquake: low-quality material use, missing support columns on upper floors, support beams across large areas resting on weak columns, narrow beams insufficiently anchored in support columns, and flexible ground floors. Unsafe modifications were made, too, like removing support columns or partition walls to create more usable square footage without adequate reinforcement. A building's movement affected those adjacent, and buildings on corners were especially vulnerable. Two examples of tragic outcomes in 1977 have direct links to poor construction and negligent regulation enforcement. The 1977 full collapse and resulting fire of Bucharest's Belvedere building (7 Brezoianu Street), constructed in 1938, killed 90 (48 identified) of its 146 residents.[16] While architects initially designed it with eleven stories, the interwar government approved an amended plan for twelve, and its builders went ahead and built fourteen.[17] In another case, an architect wrote to the state-owned construction company about Bucharest's Nestor building (2 Pictor Grigorescu Street, Stair A), which was just forty years old at the time of its partial collapse in 1977, warning that "the building did not conform to the project plans."[18]

Many in the Ceauşescu regime knew of the 1940 earthquake investigative reports and were aware that the fascist- and communist-led governments before them had largely disregarded the recommendations. On Sunday evening, two days after the 1977 earthquake, the PEC discussed the 1941 Achim report, with Ceauşescu informing members, "we must look very seriously, it seems that some [of the collapsed buildings] were seventy to eighty years old and made of old materials."[19] Within that first month after the 1977 earthquake, Romanians learned about or were reminded of the insufficient structural supports for the reinforced concrete used in interwar buildings. People knew, too, of the investigations and reports that followed the 1940 earthquake. The Securitate noted that some "experts" and other "diverse categories of Bucharest citizens" talked about how "an inventory was made in 1940 of the apartment buildings that needed repair, but World War II began and no one was concerned with their reinforcement and after that, it was forgotten to take the necessary measures."[20] Almost all the buildings that fully collapsed in 1977 were built in the interwar era using reinforced concrete without proper structural support for Romania's particular seismic risk.

The government under Gheorghe Gheorghiu-Dej also knew about yet did little to reinforce the structurally unsafe buildings. The deaths, injuries, and damage in 1977 could have been avoided if any of the post-1940 Romanian governments acted on the investigations. In a repetition of the 1940 earthquake aftermath, following 1977 a few prominent builders wrongly claimed that the technical knowledge "had not been raised in general" before 1940 for how to build for earthquake resistance.[21] Specialists at the time proved, however—and today's engineers, experts, and scholars reiterate—that the technical knowledge for designing and constructing buildings using reinforced concrete to withstand an earthquake was available. The application of it with fidelity was the problem. For the handful of buildings constructed after 1940 that collapsed in 1977, Ceauşescu blamed the individuals at the construction companies who built them, not the communist state that owned and employed them, nor the regulations it passed that should have prevented such damage.[22] In his view, individuals alone bore the burden of blame and financial responsibility for repairing the damage caused by the earthquake: "The state will not pay. We will identify who was guilty and we will dock his salary."[23]

Commissions Dispatched, Evacuations Reduced, and Residents Frustrated

Hours after daybreak the morning after the earthquake, while first responders continued the search and rescue, hundreds of experts began assessing the more than ten thousand damaged buildings across Romania, 90 percent of which were in the capital. They worked to determine the immediate structural repairs, recommend evacuations, and draft recommendation plans for both. As engineer Gheorghe Ursu, who led an assessment commission, explained, "immediately after March 4, even the following morning, all the competent professional forces mobilized, the engineer planners, builders, university professors, researchers in the field of structural engineering to evaluate the state of the buildings damaged by the earthquake."[24] The regime followed experts' lead. Two days after the earthquake, Ceauşescu told PEC members to create assessment commissions and appointed Prime Minister Mănescu to oversee them.[25] Hundreds of specialists—seismologists, geologists, students, teachers, and others—fanned out across Romania to assess the damage and draft work plans.[26] The third day after the earthquake, Ceauşescu noted,

> there's panic in the capital. This must be stopped. It is known that especially the old [apartment buildings] are damaged, but everything must be done quietly, we must explain what we will do, but gradually. We cannot just take all the people and put them in the street.[27]

Civil engineers and other experts worked without interference for just a few days before the regime took hold of the process. In Bucharest, two separate commissions assessed buildings: one for buildings five stories and taller, and the other for those shorter. These height categories for seismic stability assessments would persist through the communist and into the postcommunist eras. In 1977, Bucharest mayor Dincă managed the central Bucharest commission, which assessed the capital's taller buildings, and the eight city sector leaders, together with their local executive committee presidents, each oversaw three commissions assigned to assess shorter buildings.[28] By the third day of recovery efforts, Ceauşescu perceived "chaos" and "panic" in the capital's streets and neighborhoods from the collapsed buildings, uncertainty among residents about their safety, numerous dispatched assessment commissions, hundreds of resident evacuations, and overall confusion among experts, residents, and regime officials about the assessment, evaluation, and repair recommendation process. Ceauşescu disagreed with the experts' conclusions. As a result, state leaders intervened, drastically changed the assessment and repair protocol, and approved fewer evacuations.

Within days of the earthquake, Ceauşescu demanded that experts complete assessments before authorities approve evacuations because he thought that their recommendations had become excessive, contributing to the "chaos" and "panic." In response to one regime official's explanation for why he evacuated a state-run community meetinghouse and planned to evacuate other buildings, Ceauşescu barked:

> Not that way, dear! We must go to them and view the situation. It can't be that they want to move and just move themselves. This is not allowed! … It is not allowed to move people until we have looked! … No one is to move without approval. This must be clear! … We are not allowed to create anarchy![29]

Elena Ceauşescu interrupted him to explain that at her scientific institute, some people were moved. Mayor Dincă reassured both that no one had been moved without approval, to which Ceauşescu responded, "not even one central department can move without approval; not even those in the counties."[30] As the weeks and months of disaster response passed, Ceauşescu's obsession with the "chaos" and "panic" on the streets increased.

The US Embassy in Bucharest recognized the regime's evacuation pivot. In a telegram to Washington, DC, its staff reported that the "Romanian government has become more cautious about evacuating damaged housing."[31] Whether from panic or fear of liability, some assessment team members during those first days erred on the side of caution and overestimated damage, aware they would be held responsible if a building were to collapse. Engineer Ursu explained,

> I must say that the first stage of analysis gave some alarming conclusions. As a way to escape the liability of a specific salaried socialist, a vast majority of the buildings assessed in these first few days were declared uninhabitable. Because of this exaggeration, it is possible that it was not only a shirking of responsibility, but rather possibly a state of panic, which was present everywhere, like a pile of ruins that pushed the specialists toward a unanimous verdict.[32]

Government officers were also upset about the large number of evacuee recommendations because they did not have the labor force to assess the damage. It is not known who brought these possibly "exaggerated" evacuation numbers to

Ceaușescu's attention, but in a rash decision he intervened, annulling many of those first assessments and issuing new guidelines. Specifically, the state inspector for construction, Mircea Georgescu, reversed "previous opinions, those that were too large, putting pressure on project managers, and threatening with a specific harsh tone."[33] The regime gave new assessment guidelines in the second week after the earthquake.[34] The task was enormous, and engineering experts continued to recommend evacuations to keep people safe, give themselves time to assess, and reduce their liability in case of future damage. But Ceaușescu did not approve, unsettled by the disruption in day-to-day life the earthquake's damage and disaster response had brought. As reported by state-run media, the experts on the assessment commissions worked for only three days before Ceaușescu stopped evacuations from buildings the regime had assessed as having only minor damage.[35] In actuality, he ordered an end to almost all evacuations, regardless of the extent of the damage, and the regime evacuated fewer residents than recommended by specialists.

To further complicate its disaster response, the regime officially promised housing to all who needed it.[36] For some residents of affected buildings, the temporary or permanent housing options were untenable. The Securitate reported, "some residents [of a five-story building at risk of collapse] refused to be evacuated because they were not given comparable homes."[37] City assessment commissioners agreed with these residents and refused to evacuate them because they also believed the housing alternative to be inappropriate. In some cases, people evacuated without permission. And, assessment experts approved evacuations but did not tell the residents where they could move.[38] Many others, however, stayed in damaged buildings and waited until an assessment team arrived.

Within a month of the earthquake, the government had assessed half the damaged buildings.[39] The state's leaders would consider new assessment proposals, but they had already decided that regardless of their findings, the number of resident evacuees would be reduced from those initially proposed. Residents were less than pleased about the state's handling of affairs. A few only expressed concerns about the commissions' assessment and repair recommendations, while others defied them outright. A month after the earthquake, a group of residents refused to allow workers and an engineer to make reinforcement repairs to their building. The residents were upset that they had not been informed about the planned work, and they wanted to evacuate before repairs began. They also said that six years earlier an engineer on the commission had worked on their building's structural reinforcement, and they did not have confidence in his abilities or his current intentions. Would he assess the damage honestly, or would he say that all was fine to defend his earlier work? The local city council sent a police officer, who was unable to calm the residents. The city council asked for a delay in the reinforcement until a second commission's assessment. Ten days after the police intervened, the Securitate noted,

> the citizens have decided no longer to live in the building ... and do not have any confidence in a new reinforcement as all the walls, including those from the ground floor to the fourth floor, are crumbling from shoddy concrete, even though the building was reinforced in 1971 by three teams.[40]

The regime also reduced repair approvals in number and scope. Ceauşescu ignored the specialists' opinions if they were not in line with his own. He told the PEC that experts "have their say" but "then we will decide" and announced the new assessment guidelines, where "the number of apartments or families that would be evacuated will be reduced from the initial proposals."[41] The regime, and not scientific experts, determined who was safe to stay. The new guidelines instructed experts to calculate whether a building could "hold" its static load—that is, not crumble under its weight plus that of all its contents—disregarding additional consideration for seismic resistance. As one expert who made structural assessments during those first months said, "in other words, ignore what would happen if there would be a new earthquake tomorrow."[42]

This push to reduce evacuations to quell what Ceauşescu perceived as "chaos" and "panic" actually created significant disruption in the city and the lives of its residents, who continued to live in unsafe buildings. That spring people living in buildings with visible damage from the earthquake were clamoring for information. People complained that the assessment commissions stopped coming.[43] In some cases when the commissions did arrive, they "were attacked by the citizens they visited."[44] The Securitate noted "a state of dissatisfaction among the citizens who wait to be evacuated; they waste time waiting for commissions and are alarmed by the incompetence of one commission or the other."[45] People felt that the assessments were incomplete or incorrect: "Thousands of citizens after they were assessed, requested a new commission to come, because the first one did not see all the cracks and within time another appeared."[46] Some tried to skip the assessment queue with a bribe: "Thousands of citizens accosted commission members on the streets, offering small presents for their help, a carton of Kents, for example, and were offended when refused."[47] The Securitate noted, "residents and their families were ready to go to the State Council and request that the decisions of the head of state be applied and not delayed, as was the case until now." And they "threatened to present a memorandum" to the Central Committee of the Romanian Communist Party (CCRCP) with complaints that their building had "severe cracks" yet the commission that assessed it "did not give a definitive situational account for the future of the building."[48] Some pleaded for their buildings to be assessed and repaired while at the same time threatening to bypass the local authorities and appeal directly to Ceauşescu and the Central Committee of the RCP.[49]

People complained about no-show commissions and others that arrived without experts. The Securitate recorded, "a frequent comment was that some of the assessment commissions are made up of people who are not specialists and as a result cannot answer citizens' questions nor are they able to reassure them."[50] Residents complained when commissions contradicted previous evaluations or gave bad advice:

> A university professor recommended that evacuees return to their building, yet to walk carefully on the stairs and in their apartments so as not to produce vibrations. These recommendations created discontent among the tenants; one threatened to go to the General Secretary.[51]

One resident reported to the Securitate that a second engineer who came to assess their building had "an irreverent attitude toward the citizens who suffered, refusing to hear the details of the damage to their homes, and did not advise them how to proceed. These attitudes created some tension among the residents."[52] Others were upset that different commissions' conclusions conflicted with one another. Residents of a nine-story apartment building were "in a state of panic because the building experienced severe structural resistance damage and the three commissions that came to assess had different opinions and did not order an evacuation."[53] Some residents were reluctant to evacuate. The Securitate noted a man who said he would "only move from the apartment building on direct orders from the party and state leadership." His anger was brought on when an assessment commissioner told him and all the residents in his building, "if you do not want to die, you should move to another building."[54]

Not only did the regime allow people to live in buildings known to be structurally unsafe, but it admitted that some reinforcement work was insufficient. Just five days after the earthquake, experts had assessed an apartment building with shops on its ground floor as at "imminent risk of collapse," yet the regime sent workers to continue with cleanup so as not to produce panic. Experts conducted a second analysis of the same building, which confirmed the first evaluation, reporting that "the building was leaning on all the metal structures and could collapse at any moment." Even with this information, the regime allowed workers to remain in the building for several hours (before allowing them to quit early and move to a different worksite until the building's reinforcement work was completed).[55]

The recurring theme of the Ceaușescu regime's disaster response was that people often had to fend for themselves. Some waited for assessment commissions to recommend whether to evacuate, and others went ahead and evacuated without approval. In one case, within two days of the earthquake, 90 percent of the residents of one building left without approval. Some moved their belongings to the street or their building's courtyard, contributing to what Ceaușescu saw as "chaos."[56] Others evacuated themselves before an assessment commission arrived because their building had "visible cracks on its interior and exterior [walls] and were restless and worried" because their building's structural safety had not yet been determined. In another example, all residents of one building and three-quarters of four buildings in a five-building apartment complex evacuated because three of their neighbors, "specialists" on an assessment commission, evacuated themselves and their families.[57]

That spring, in addition to being upset about how the regime was handling the assessment, reinforcement, and repair process, people questioned who was responsible for their building's resistance failures. In one case, a group complained to the building's supervisor, who then reported to the Securitate, that "the state should give money back to them because their homes were not built correctly." The Securitate noted that assessment specialists "affirmed that the earthquake damage was due to builders' negligence and superficial work, as well as a lack of oversight from local leaders."[58] Residents living in damaged buildings sought to ensure that those people responsible for the poor construction would not be part of the 1977 assessment commission because they might alter their conclusions to shirk their liability.[59]

The Demolition Versus Repair Debate

In mid-April, after having assessed thousands of buildings in just over a month, specialists recommended 350 buildings in Bucharest for the most critical repairs, determining that they "needed reinforcement in the first urgency category." Engineers immediately began repairs on half of them. The experts placed buildings in categories based on temporal assessments, a kind of repair triage system. These were not categories based on specific structural needs. The first urgency category (U1) included those buildings determined to be so structurally unstable that they would collapse in the next strong earthquake. Experts assigned about two thousand five-story and taller buildings to lower categories that needed less urgent (but no less essential) reinforcement, and thousands of smaller homes that needed minor repairs.[60] The repair job was immense, spurring a debate between regime leaders and specialists about which course of disaster response was best: demolish the buildings or make the needed structural reinforcement repairs. On the ground, this looked like "a dull and uneven fight" of technical specialists against regime apparatchiks:

> On one side were the engineers, responsible for the reinforcement work of the damaged apartment buildings trying to defend their projects, to confer acceptable antiseismic structural resistance, far from absolute certainty, but consistent with commonsense engineering. And, on the other side, were the Bosses ... and some other specialists of high competence, obliged to support the view from above and looking for ridiculous arguments, which was clear they did not believe, yet denying their own conclusions of a week or a month before.[61]

The debate was a test of specialists' and responsible actors' ability to align with the regime's expectations in the face of significant public safety risks.

That spring, Ceaușescu consistently supported repair over demolition and, at times, even extensive reinforcement repairs over cosmetic ones. He had no patience for scientific discussions or nuanced debate about the best ways to handle structural repairs or building safety, saying, "we are not sitting here to have a theoretical debate, we must go and repair."[62] He opened the March 30 PEC meeting saying, "know that we are demolishing all of Bucharest, it will be nice," but he was not advocating for the city's demolition, rather he was commenting on its chaotic state.[63] Until July he consistently supported repair even though in March he publicly announced his plans to build the Civic Center Project and its House of the People (Appendix A). That first week after the earthquake, Ceauşescu agreed that there were general and seismic reinforcement repair needs, telling the PEC, "I think that we must urgently move to reinforce and start the repairs, as was discussed yesterday, to move urgently to reinforce."[64] He was not sure whether Romania would (or even could) experience another strong earthquake, but he agreed on the need to reinforce buildings for seismic stability, regardless, rambling,

> it is true, comrades, it is fair to say, that this earthquake, better said of that intensity, will not be repeated, yet others have said it will be repeated … there are many discussions, some claim that after an earthquake that strong, which happened in Romania, other earthquakes will not repeat; others say that it can be repeated, but not at that intensity.… Therefore we must immediately take measures to reinforce the damaged buildings.… This week work must happen day and night to realize the repairs necessary, the reinforcement of those.[65]

Early on he even agreed that experts must lead the efforts, proclaiming,

> each engineer must work on reinforcement. It is difficult to say here what there is to be done, and what measures for reinforcement to take. Every engineer must go look at every house and determine what sort of reinforcement to make.[66]

Ceauşescu insisted that damaged buildings must be assessed several times before evacuation orders or repair plans could be issued and, moreover, that he alone could authorize demolitions: "I gave a few communications and I took some very firm measures. … I review three times and on the fourth comes my approval and only after that [can they demolish]."[67] Others in the regime echoed his preference for repairs. Just two days after the earthquake, Gheorghe Cioară, PEC member and government vice minister, reported that fifteen assessment commissions working in Bucharest had

> identified some apartments that are necessary to evacuate immediately and those that don't, and what needs immediate demolition, and what does not. At the same time, it is clear what immediate measures are necessary and what could be reinforced because we do not want to demolish everything.[68]

Ceauşescu did not rule out demolition, but initially, he actively worked against a mindset that it was the best solution. Some PEC members argued for demolitions or more time to assess and repair. Mayor Dincă tried to convince Ceauşescu to demolish some damaged buildings, arguing,

> Comrade Secretary General, we have some more buildings, which I will present to you, that must be demolished. I report to you that we have in Saint Vineri Square some homes that are very damaged. They need serious repairs and are in an advanced state of degradation. Of these, we have one apartment building with sixty-seven families that is completely cracked, but we still have not evacuated the families. The commission has seen it and they drafted the act to request its demolition. Otherwise, [if not demolished], it will enter into systematization. If it is reinforced it will last only a year or a year and a half and require serious reinforcements.[69]

Ceauşescu railed against those who supported demolition, like Dincă, as a first response, explaining, "we cannot have people arrive with the idea to place explosives and to blow [the damaged buildings] up." He then responded to the mention that there had been requests for demolitions: "Some say, it is true, that we should destroy Bucharest. This is what people we have! Yes, comrades, indeed these people must be removed from their work responsibilities because they represent a dangerous line of thinking."[70] Ceauşescu informed Valentin Drozdenko (b. 1924), Soviet ambassador to Romania, about the disaster response's first month, saying,

> of course, there are many homes that have suffered a lot; the repairs of those will take a long time. We are starting to reinforce as many of them as possible so we are not obliged to demolish them. This is especially true for the large buildings.

Drozdenko responded, "there is much work to do." Ceauşescu continued, "also, there are serious problems because we must make [the buildings] resistant." He described the extent of the work and admitted that it was moving slowly: "Practically, it will take all year and some of the next."[71]

Ceauşescu continued to resist demolition. On April 6, a month after the earthquake, he announced to the PEC that "nothing will be demolished without approval—I am referring to the apartment buildings."[72] The regime published an announcement that any demolition of buildings damaged by the earthquake required state approval:

> the Political Executive Committee of the Central Committee of the Romanian Communist Party decided the following: the party and state departments and organizations will continue to act with all the force necessary in the economy, construction, and all domains of economic and

> social life for the earthquake relief efforts and for ensuring the realizations of the best conditions foreseen in the country's development plan, and continue to raise the level of material and spiritual living standards of those who work. It will strictly adhere to decisions previously made, which prohibit the demolition, without special approval, of residential buildings damaged by the earthquake.[73]

Despite Ceauşescu's initial views, many specialists thought that demolition was more cost effective than reinforcement repairs.[74] The Securitate noted, "specialists conclude that the amount of money needed to reinforce [a] building is so much that it is more economical to demolish and construct a new building." An architect and manager at the systematization department said, "regardless of the methods and solutions used to reinforce the damaged apartment buildings, it cannot be guaranteed that they will not collapse in the event of a possible earthquake. To be economical, they should be demolished and rebuilt."[75] Another argument for demolitions as the best course of action was that Romania did not have access to the materials to complete retrofitting repairs. The Securitate reported,

> other specialists from the institute's workforce said that reinforcing buildings is much more expensive than demolishing them and rebuilding from scratch. We do not have a tradition of manufacturing suitable materials for reinforcements, which are made to order, almost specific for each property, and when they are unavailable we improvise with inappropriate materials.[76]

In light of Ceauşescu's initial position against demolition, some who supported it were hesitant to say so.[77] By that May, Minister of the Interior Theodor Coman supported demolition, citing Romanian, American, and Soviet experts who had determined that the necessary research had not been done, that repairs would be costly in time and effort, and that demolition was best. He added that "some Romanian seismologists had the opinion that some areas of the capital most affected by the earthquake … should be completely demolished since the thirty to forty buildings would hinder the optimal arrangement of the respective thoroughfares."[78] He used American and Soviet conclusions to support his position that the repairs made during the first two months following the earthquake were superficial, cited experts' demolition recommendation, and added that time was needed, saying, "the reinforcement problem is more important and more difficult than new construction and can only be resolved after a detailed study of a building's strength, which requires at least three months of work."[79]

Government leaders and construction experts also considered the urban planning projects that had been set in motion across communities in Romania. Long before the 1977 earthquake, during the first year of his reign, Ceauşescu had introduced systematization (*sistemizatizare*), the national territorial planning policy that aimed to urbanize, modernize, and reconstruct rural Romania by moving people from villages to towns, locating factories in smaller towns, and demolishing homes for modernization efforts.[80] After the 1977 earthquake, systematization expanded to towns and cities, including Bucharest. Regime leaders and directors, alongside architects, managed the policy, leaving construction specialists to implement it.

That spring, neither side won the debate. Either solution—demolition or reinforcement repairs—would require resident evacuations, months or even years of work, and significant expenses in time, labor, and materials. Since thousands of buildings needed some intervention, Ceauşescu ultimately found both options untenable. In July 1977 he privately proposed a third solution for the damaged buildings: end assessments and hasten repairs (Chapter 6). If experts assessed fewer buildings as uninhabitable then fewer people would require evacuation and, as his logic went, fewer buildings would need reinforcement repairs. Ceauşescu did not use economics as justification for the decision; rather, he was fed up with the ongoing "chaos" and "panic" he saw during the four months of disaster response.

The First Disaster Response Deadline

The regime reduced the possible number of assessments by establishing May 1 as the deadline for their completion.[81] The regime chose this day, the socialist Labor Day, so that "chaos" and "panic" would be removed in time for the holiday's large-scale celebrations and orchestrated, mandatory parades all across Romania, with the largest in Bucharest (see Figure 5.1). Even though the regime had removed the piles of rubble from the collapsed building sites, the disruption from evacuations and the repairs underway meant that in Ceauşescu's eyes Bucharest was still a mess. Declaring that a building did not need to be evacuated and stalling repairs were easy ways to limit disruption, clearing the streets for the May 1 celebrations.

As the May 1 deadline approached, experts worked urgently.

> From the first stage of the retrofitting, project managers, experienced construction engineers, and technical staff from the university were appointed. Everything had to be done on the fly, there was no time for in-depth research, the head of state had said in a speech in March, that … by May 1, 1977, the traces of the earthquake must be completely liquidated.[82]

Superficial repairs became more common. Experts were concerned about the work's quality, with the Securitate noting that commissions were "plastering walls, leading to the high risk of concealing major damage, and limiting the commissions' ability to assess the actual situation of the buildings and make structural reinforcement repairs."[83] An architect in one construction group said that his team made "inappropriate" building repairs and plastered over cracks in the walls instead of introducing a filler that would provide seismic

Figure 5.1 Nicolae Ceauşescu (front row, fifth from left) and regime leaders stand during the May 1 celebrations in Bucharest, 1966.

Source: Fototeca online a comunismului românesc, BA166, Cota 26/1977.

resistance.[84] Engineers and other experts scrambled, "but the deadline was absurd, impossible to meet."[85] Most made what assessments they could. During a meeting just five days before the May 1 deadline, Mayor Dincă stated that the regime wanted all buildings evaluated by that time so "that people can live in good conditions. Therefore, we are requesting a helping hand from planners and builders so we can apply ourselves to reach the best solutions."[86] He too, recognized that the May 1 deadline was impossible to meet and asked for more time and human resources, appealing to Ceauşescu and the PEC: "We need eighty more days just for the assessment, and after that … we can move to reinforcement. And, I would like to revive this commission to the size it was before with the same rate of work for us to complete all of the assessments in the capital."[87] The stenographers did not record Ceauşescu's response, but he did not grant the request. Experts did not meet that May 1 deadline.

Conclusion

During the spring months of the 1977 earthquake disaster recovery, the regime undertook several activities to assess the damage. It reviewed investigations made after the 1940 earthquake (and collapse of the Carlton building), determining that precommunist governments had recommended, but not completed, significant seismic stability repairs for a majority of those buildings that collapsed in 1977. The Carlton building's collapse and tragic death toll highlighted, to both the fascist regime of the 1940s and the communist regimes that followed, the risk when building with reinforced concrete without adequate earthquake resistance.[88] The regime learned how interwar builders cut corners and postwar Romanian governments looked the other way, allowing for construction that did not use seismic structural resistance knowledge or follow contemporary building codes.

Scientific experts began an assessment process to determine the extent of the structural damage to thousands of buildings, almost all in Bucharest, to approve evacuations, make plans, and begin reinforcement repairs. Experts worked diligently to complete as many assessment plans as possible. They were charged, too, with recommending resident evacuations because many buildings were unsafe for habitation. Initially, they worked without interference. Yet, within just a few days, Ceauşescu complained about the "chaos" and "panic" he saw in the capital among experts, officials, and residents. Just one week after the earthquake the regime took hold of the process, with Ceauşescu challenging experts' conclusions and recommendations about evacuations and repairs. Much like after the 1940 earthquake, the engineering experts working in 1977 never finished assessing or reinforcing all the buildings damaged. During the four months following the earthquake, Ceauşescu was extremely concerned about the sight of the damaged buildings and the repair teams, as well as the many dispatched assessment commissions, hundreds of resident evacuations, and ongoing confusion about the assessment and repair processes, and so he ordered Bucharest's streets clear ahead of the May 1 celebrations.

From the initial days after the earthquake until July 1977, Ceauşescu supported repairs over demolition. Yet, his rhetoric was inconsistent. At the same time that he claimed he did not want to demolish the buildings damaged by the earthquake, he resurrected plans for the Civic Center, his large-scale urban planning project with its gigantic House of the People, with the earthquake a clear catalyst for this decision. Many of the houses, apartment buildings, churches, monasteries, and other buildings razed for the project's site—almost 25 percent of the city—had not been damaged by the 1977 earthquake. Ceauşescu's preference for repairs over demolition held only for those buildings damaged by the earthquake and not for those in the way of his modernization plans.

The agency and authority Ceauşescu gave to engineering and construction experts the first week after the 1977 earthquake eroded quickly to nil. In the weeks and months that followed Ceauşescu changed his opinion drastically about the need to reinforce and repair the damaged buildings. Over the first several months of the disaster response, the regime retreated from committing to or engaging with substantial repairs and then reduced and ultimately stopped evacuations. These incremental steps ended that July, when Ceauşescu ordered the end to all assessments and a quick completion of repairs underway to hundreds, if not thousands, of buildings (Chapter 6). Ultimately, he chose the path of least resistance. As a result, Romanians continued to live in almost two hundred thousand damaged buildings, with more than one-quarter considered by specialists to be uninhabitable, almost one thousand of them expected to experience significant damage in the next major earthquake, and almost four hundred expected to collapse when such an event occurs.[89]

Notes

1 World Bank, *Report and Recommendation*, 13.
2 Secţia Cancelarie, 21/1977, 12 rev., FCCPCR, ANR.
3 E.-S. Georgescu, "The Collapse of Carlton Building," 60.
4 Vlad, "Causes and Effects," 124–25.
5 Emil-Sever Georgescu, email message to author, May 1, 2016.
6 Petrovici, "The 10 November 1940," 90–91; Vlad, "Causes and Effects," 126.
7 Vlad, "Causes and Effects," 123–24.
8 Calotescu, Neagu, and Lungu, "Before and after November 10th," 44–45.
9 Petrovici, "The 10 November 1940," 89.
10 Calotescu, Neagu, and Lungu, "Before and after November 10th," 45, 47.
11 Petrovici, "Architecture Face to Face"; Vlad, "Causes and Effects," 126.
12 Vlad, "Causes and Effects," 116–18.
13 Petrovici, "The 10 November 1940," 90–91.
14 Vlad, "Causes and Effects," 124–25; Berg, "Earthquake in Romania," 9–10.
15 Berg, "Earthquake in Romania," 11; World Bank, *Report and Recommendation*, 17.
16 Antonios Pomonis, email message to author, April 11, 2016.
17 Petrovici, "The 10 November 1940," 93; D.14.800, v. 1, 217–18 rev., DB, CNSAS.
18 D.13.339, v. 37, 307, DB, CNSAS.
19 Secţia Cancelarie, 21/1977, 12 rev., FCCPCR, ANR.
20 D.13.339, v. 37, 306, DB, CNSAS.
21 Petrovici, "The 10 November 1940."
22 Secţia Cancelarie, 35/1977, 3 rev., FCCPCR, ANR.
23 Secţia Cancelarie, 78/1977, 42 rev.
24 *Povestea Vorbei*, no. 195, 1, RFE/RL Romanian Broadcasting Department records, box 3811, folder 2, HILA.
25 Secţia Cancelarie, 20/1977, 14 rev., FCCPCR, ANR.
26 Secţia Cancelarie, 27/1977, 54 rev.; Secţia Cancelarie, 21/1977, 206.
27 Secţia Cancelarie, 23/1977, 13 rev.
28 World Bank, *Report and Recommendation*, 14.
29 This response was addressed to Aurel Duma, member of the CCRCP's Secretariat, March 6, 1977; see Secţia Cancelarie, 20/1977, 8, FCCPCR, ANR.
30 Secţia Cancelarie, 20/1977, 8–8 rev.
31 ActionAid Agency for International Development, "Earthquake: Sitrep 4 March 8."
32 *Povestea Vorbei*, no. 195, 1.
33 *Povestea Vorbei*, no. 195, 2.
34 Boia, *Strania istorie*, 214; Secţia Cancelarie, 35/1977, 43, FCCPCR, ANR; Secţia Cancelarie, 42/1977.
35 James Edwards to Ralph Walter, Nathan Kingsley, and Thomas Bodin, memo, March 9, 1977, RFE/RL Romanian Broadcasting Department records, miscellany, box 1858, HILA.
36 "Romania Earthquake," RFE Lyon to Washington, DC memo, March 16, 1977, RFE/RL corporate records, box 329, folder 15, HILA.

37 D.13.339, v. 37, 72, DB, CNSAS.
38 D.12.639, v. 15, 142, DB, CNSAS.
39 Secţia Cancelarie, 48/1977, 21 rev., FCCPCR, ANR.
40 D.13.339, v. 37, 336 rev. and 337, DB, CNSAS.
41 Secţia Cancelarie, 27/1977, 24, FCCPCR, ANR.
42 *Povestea Vorbei*, no. 195, 1.
43 D.12.639, v. 15, 107, 142 rev., and 178, DB, CNSAS; D.13.339, v. 37, 73 rev., 162 rev., and 204, DB, CNSAS.
44 *Povestea Vorbei*, no. 195, 4; D.12.639, v. 15, 7 and 86, DB, CNSAS; D.13.339, v. 37, 73, 103, 307, and 336, DB, CNSAS.
45 D.12.639, v. 15, 143, DB, CNSAS; D.13.339, v. 37, 163, DB, CNSAS.
46 D.12.639, v. 15, 7 and 86, DB, CNSAS; D.13.339, v. 37, 73, 103, 307, and 336, DB, CNSAS.
47 *Povestea Vorbei*, no. 195, 4.
48 D.13.339, v. 37, 73 rev., DB, CNSAS.
49 D.12.639, v. 15, 7 and 143, DB, CNSAS; D.13.339, v. 37, 104, 204–5, 213, 326, 349 rev., 365 rev., 366, 373, and 373 rev., DB, CNSAS.
50 D.12.639, v. 15, 8 rev., DB, CNSAS.
51 D.13.339, v. 37, 327 rev., DB, CNSAS.
52 D.13.339, v. 37, 327 rev.
53 D.13.339, v. 37, 74.
54 D.12.639, v. 15, 8, DB, CNSAS.
55 D.13.339, v. 37, 94–94 rev., DB, CNSAS.
56 D.13.339, v. 37, 73 rev. and 213.
57 D.13.339, v. 37, 73 rev. and 249.
58 D.13.339, v. 37, 73 rev. and 249.
59 D.12.639, v. 15, 142 rev., DB, CNSAS, DB; D.13.339, v. 37, 162 rev., DB, CNSAS.
60 *Povestea Vorbei*, no. 195, 2.
61 *Povestea Vorbei*, no. 195, 3.
62 Deletant, 37.
63 Secţia Cancelarie, 42/1977, 2, FCCPCR, ANR.
64 Secţia Cancelarie, 27/1977, 54 rev.
65 Secţia Cancelarie, 27/1977, 54 rev.
66 Secţia Cancelarie, 27/1977, 35 rev.
67 Secţia Cancelarie, 27/1977, 35.
68 Secţia Cancelarie, 21/1977; Dobre, *Membrii C.C. al P.C.R.*, 152–54.
69 Secţia Cancelarie, 48/1977, 32–33, FCCPCR, ANR.
70 Secţia Cancelarie, 21/1977, 51 rev.
71 Secţia Cancelarie, 48/1977, 32–33.
72 Secţia Cancelarie, 48/1977, 25.
73 Secţia Cancelarie, 48/1977, 3 rev.
74 D.13.339, v. 37, 365 and 386 rev., DB, CNSAS.
75 D.13.339, v. 37, 349 and 365.
76 D.13.339, v. 37, 365.
77 D.13.339, v. 37, 365 rev.
78 Secţia Administrativ-Politică, 2/1977, 43 rev., FCCPCR, ANR.
79 Secţia Administrativ-Politică, 2/1977, 43 rev.
80 Deletant, *Ceauşescu and the Securitate*, 300.
81 Secţia Cancelarie, 27/1977, 6, FCCPCR, ANR.
82 Tofan, *Antologia Radio Europa Liberă*, 84.
83 D.12.639, v. 15, 174, DB, CNSAS.
84 D.13.339, v. 37, 365 rev., DB, CNSAS.

85 *Povestea Vorbei*, no. 195.
86 Secţia Economică, 59/1977, 27, FCCPCR, ANR.
87 Secţia Economică, 59/1977, 27.
88 Georgescu and Pomonis, “Building Damage.”
89 Secţia Cancelarie, 35/1977, 92, FCCPCR, ANR.

6 Ending the Disaster Response

The July Cessation Order

Exactly four months after the 1977 earthquake, during a private meeting on July 4, Nicolae Ceauşescu essentially ended the disaster response when he ordered all structural assessments of damaged buildings in Bucharest to stop completely, the scope of drafted repair plans to be reduced, and any repairs underway to be finished by a six-week deadline, pinned to yet another important state holiday.[1] Ceauşescu explained to regime officials, construction specialists, and engineers at the meeting, "comrades, what must be done immediately is first, to stop the [Bucharest] sector-level assessment commissions. Today, I ordered police and prosecutors to arrest and imprison anyone who enters people's homes: no one has the right to enter anyone's home! ... We proceed immediately to repair homes where people live or have evacuated and, as was decided in March, require every citizen to repair his own home."[2]

During the first phase of disaster response, over four months specialists and regime officials assessed the damaged buildings, made evacuation recommendations, and began repairs, while regime officials investigated why so many buildings collapsed, debated the merits of demolition versus repair, and reduced initial evacuation recommendations for homes and workplaces. Ultimately, Ceauşescu chose neither to demolish nor to repair: he chose to end the work, essentially stopping all that needed to be done. The July cessation order established a deadline of August 23 to complete repairs and clean worksites in time for parades and festivities in the streets to celebrate Liberation Day, one of Romania's important holidays. Ceauşescu's strongest motivation was simply to eliminate the "chaos" and "panic" he perceived in the disaster response all across the capital. Specialists and engineers met the order with trepidation, concerned about public safety and their liability in case of insufficient repairs. As he did throughout the disaster response, however, Ceauşescu disregarded the scientific experts' opinions and recommendations. As a direct result of the order, today, almost one thousand buildings in Bucharest are at risk of serious damage, with a third expected to collapse in Romania's next major earthquake.[3]

DOI: 10.4324/9781032632933-7

The July 4 Meeting Ending the Disaster Response

During the private July 4 meeting, before announcing the order, Ceauşescu asked Bucharest's eight sector leaders to report on the assessment, evacuations, and repair progress of damaged buildings in their jurisdictions. Each reported practically the same number of assessments, around two thousand. Considering that the earthquake had caused some damage in six of the city's eight sectors, but significant damage in two, this uniformity suggests that leaders were unsure what Ceauşescu wanted to hear. Did he expect larger or smaller numbers? How would he respond if one sector reported more than another? It was difficult to navigate his whims. Ceauşescu also asked whether high school or university construction and engineering students had participated in the assessments. Again, specialists faced a dilemma:

> How was it best to respond? If you said the high school or university students did not participate, you could be accused of being guilty of a lack of confidence in the youth, the new, dynamic, fresh elements. If you said that they did participate, you could be accused of lacking competence, having an "adventurist" character, or allowing some runt to decide the fate of a building.[4]

Uniformity in answers was one solution to the problem of navigating Ceauşescu's unpredictable expectations.

The Bucharest sector leaders participated in this balancing act, reporting while gauging Ceauşescu's expectations. The exchange between him and Adriana Moraru, leader of one of the two Bucharest sectors with the most earthquake damage, is illustrative of the meeting's tone, the regime's concerns, and Ceauşescu's approach to the disaster response and the experts key to it:

Adriana Moraru:	We organized [assessment] commissions of two to three specialists. They evaluated 2,480 damaged buildings.
Nicolae Ceauşescu:	Who permitted you to evaluate 2,400 buildings?
AM:	Comrade Secretary General, the people came and asked.
NC:	What people came?
AM:	Citizens, they asked us to see their buildings.
NC:	How many citizens did you evacuate from the buildings, do you know?
AM:	Us, evacuate?
NC:	Yes, yes.
AM:	No …
NC:	Your commissions, not you. You did not go to talk to citizens, you sent boyars. The commissions that went, how many did they evacuate?

AM: They were not evacuated, Comrade Secretary …
NC: No, fine, I will show you how many were evacuated.[5]

Ceauşescu challenged Moraru's claim that commissions in her sector assessed more than 2,400 buildings and outright told her she did not know that evacuations had occurred. His focus on how many people were evacuated had more to do with his concern about reducing disruption caused by the displaced than with any concern for their housing situation or safety. He accused her of not going herself to assess the damage and claimed that she sent "boyars," a reference to fourteenth- and fifteenth-century Romanian landowners and eighteenth- and nineteenth-century Ottoman Phanariotes dispatched to govern the Romanian principalities of Wallachia and Moldova for the sultan. His not-so-subtle comparison of her to the landowning ruling class and the foreign Christians working for the Ottomans was a clear criticism of her leadership. The confusion among the leaders at the July 4 meeting—from Ceauşescu down through the local Bucharest sector administrators—about the assessment and repair process is palpable in this exchange with Moraru.

Ceauşescu then addressed Bucharest mayor Ion Dincă: "The building engineers arrived and the tenants complained. Yes, Comrade Dincă, we shrug our shoulders because we left the capital a mess and we did not lead it. We destroyed it. We did what the earthquake did not do."[6] Ceauşescu accused Dincă of dismissing the chaos and panic of the city and pointed to the mayor's fatigue from the months of disaster response. Over the previous four months, Mayor Dincă had changed his stance about the best course of action. Initially, he believed demolition to be best, but by July he concurred with Ceauşescu that it would be best to stop assessments and hasten repairs. He took responsibility for the chaos Ceauşescu perceived in the capital, explaining, "as for the repairs, of course, a series of initial mistakes were committed, which in our opinion were because of the need to reinforce the entire building as opposed to [making] localized repairs."[7] That he fell in line with Ceauşescu was expected. Dincă was with the Ceauşescu regime until its very end and was part of Elena Ceauşescu's "personal court of hagiographers."[8] At the start of the July 4 meeting, Dincă told Ceauşescu,

> I assure you, Comrade Secretary General, you sir personally, and the party and state leaders, that I understand well your criticisms of me and the bitterness you have expressed because Bucharest looks more desolate now than immediately after the earthquake…. Reinforcement? What is that? Repairs are what they are and what must be done.[9]

Dincă echoed Ceauşescu's disregard for expert opinion, claiming that some planners and civil engineers in Project Bucharest (the city's urban planning department) and other departments exaggerated what was necessary, did not base their recommendations on structural assessment calculations, and extrapolated recommendations from other buildings' findings.[10] In line with Ceauşescu, Dincă expressed a belief that many nonspecialists in the regime held: that specialists had exaggerated the need for "extensive" (i.e., structural reinforcing) repairs.

As he had in March, demanding cleanup in time for the May 1 holiday, Ceauşescu gave a deadline of one of Romania's important holidays so that the city would be clear for celebrations. In July he set the deadline for Liberation Day, celebrated on August 23, which commemorates the 1944 coup that removed Antonescu's regime. Romanians observed August 23 with mandated parade participation, with the largest parade held in Bucharest for regime leaders arrayed in viewing stands (see Figures 6.1 and 6.2). Many earthquake-damaged buildings sat along the same boulevards where parades would march, for which Ceauşescu demanded: "The capital had to be clean, without scaffolding and freshly whitewashed."[11] In March Ceauşescu had only asked for repair work to hasten and evacuations to be reduced by May 1. By contrast, in July he demanded that by August 23 construction and engineering experts reduce repair recommendations already drafted, not begin any new assessment plans, and quickly finish all repair work underway, prioritizing those with the most publicly visible damage.

Figure 6.1 Celebrating the August 23 holiday in Aviators' Square, Bucharest, 1977.

Source: Fototeca online a comunismului românesc, BA167, Cota 167/1977.

Figure 6.2 Parade through Aviators' Square, Bucharest, to celebrate the August 23 holiday, 1977.

Source: Fototeca online a comunismului românesc, BA166, Cota 166/1977.

Ceauşescu's Motivation and Disregard for Specialists

During the July 4 meeting, Ceauşescu did not justify his decision. He was not motivated by the disaster response cost or specialists' abilities to assess, plan, and make repairs. Rather, Ceauşescu needed to stop the "chaos" and "panic" he perceived. He saw the regime's disaster response as having created more disruption than the earthquake did. When he introduced the July cessation order Ceauşescu explained that it was necessary because the assessments in Bucharest had "led to an increase in panic, it created chaos inadmissible in the capital."[12] He anticipated, or wanted, a quick disaster response, but the assessments, plans, and repairs for thousands of buildings required more time. While workers had cleared the rubble from the collapsed buildings quickly, the assessments and repairs were progressing slowly, contributing to the disorder he saw, disturbing and unsettling him. He told the Political Executive Committee (PEC): "We agreed to start the repairs, to stabilize, and to repair; and within a few months to complete the repairs, not for you to demolish the capital!"[13] The assessment commissions deployed in Bucharest were daily reminders of the safety risks that remained and the reinforcement work that needed to be done. If those disappeared from public view, and if discussion within the regime about them stopped, then the risk disappeared as well. From his point of view, the various agencies that made assessments and their conflicting recommendations about whether to evacuate people, close factories or shops, or demolish buildings all contributed to this "chaos."[14] Ceauşescu complained that the assessment commissions had determined "dozens" of apartment buildings to be uninhabitable and many stores unsafe for business "without one justification, except simply a false concept and irresponsibility on the part of the Central Commission that approved this work." He believed that those closures and evacuations contributed to the "chaos," explaining, "with dozens of buildings out of service we found ourselves in a worse situation than that caused by the earthquake."[15]

An overarching, undesirable, and destructive quality shared by Nicolae and Elena Ceauşescu was their mistrust and disregard for specialists and scientific experts. Well into the second decade of the Ceauşescu regime, the ruling couple considered themselves to be the ultimate authorities on practically all things. Nicolae gave himself and approved for himself grand designations as "Genius of the Carpathians," "Architect," "Builder," "Designer," and "Architect and Constructor of the Country." There were scores of others. He considered himself the nation's patriarch, the person who had crafted Romania's version of socialism, birthed its modernism, and built its industrial state. And, he regarded with suspicion those who held actual credentials. Elena Ceauşescu, too, promoted an inflated self-image as the nation's matriarch and top scientist. She left school at fourteen yet wrote a thesis on polymer chemistry, for which the University of Bucharest granted her a PhD. Most agree that she extracted this degree and the honorary degrees she held, and published papers by ghostwriters.[16] During the July meeting, Nicolae complained to PEC members that they put the disaster response management in the "hands of professors" and in "the hands of specialists," explaining, "I do not blame our comrade planners and comrade architects for having differing opinions but instead we must seriously discuss with those from construction and with our architects what it means, this thinking, this faulty thinking, which must not be accepted."[17]

Ceauşescu often invited specialists to report to himself yet used those opportunities to instruct the experts on how they were wrong. During the 1977 disaster response, he told civil engineers that he disapproved when they removed surface plaster from columns to inspect for damage, a routine diagnostic procedure.[18] During a long diatribe with no nuance, he asked them, "If there is one crack in a column the whole apartment building goes down? Where in the world does that happen?"[19] Ceauşescu even took it onto himself to examine cracks and fissures in a damaged building and concluded that they did not negatively affect its resistance. He claimed that the reinforcement plan for the severely damaged building of the Scala Café and Movie Theater was incorrect. After the earthquake, the theater and café continued to operate until an assessment commission determined the building to be unsafe, to which Ceauşescu countered, "[the supports] were made thirty or fifty years ago and have been fine. We have not proposed to ourselves here to remake all its supports, which is the architects' and planners' idea. This would mean that we should change all of Bucharest. Where did this concept come from!?"[20]

The ruling couple's disregard for specialists' knowledge and informed opinions was not a new phenomenon after the 1977 earthquake, but the need for experts and specialists to evaluate the safety of tens of thousands of buildings across Romania brought their disregard for experts to the surface. Earlier that spring Ceauşescu essentially said that all expert opinions about a building's structural integrity should be ignored and that no building would be demolished because an expert said it was unsafe. He told the PEC:

> Those who arrive with proposals to demolish must be kindly sent home and must not reenter into the commission regardless of whether they're a professor or whomever they may be. We do not need someone to come and impress us with their expertise. This is how we will act to solve the problems: simply, without complications, and without all of this![21]

Ceauşescu's remarks are illustrative of his regime's disaster response: dismiss specialists' recommendations if inconvenient to his vision, in favor of sanitized, easy, superficial, short-term fixes. In the case of the July cessation order, this response established the dangerous, significant public risk that remains in Bucharest to this day.

Liability and Confusion with the July Cessation Order

The July cessation order was a directive from Ceauşescu for the specialists assessing the buildings, drafting plans, recommending evacuations, and making repairs. It was not a law, a presidential decree, or a public announcement. Romanians did not have access to the text of the order. They read in the state-run newspaper that disaster recovery would finish in time for the August 23 holiday, but not under what conditions. On July 8, four days after Ceauşescu announced the cessation order, the regime communicated its directives via telex to civil engineers and other specialists in two construction agencies, the Inspector General's Office for Construction and the National Research Institute for Construction.[22] The telexed order stated that "the reinforcement work for buildings damaged by the March 4, 1977, earthquake will bear in mind the strict adherence only to damage reinforcement and for all to carry out only absolutely necessary repairs" and clarified these repairs must "proceed at a sustained pace." When the supervisors responsible for the assessment commissions and repair teams received the cessation order, they immediately transmitted it to the civil engineers and other specialists they managed: stop assessments and carry out only "absolutely necessary repairs," prioritizing those for highly visible damage.[23] "Absolutely necessary" referred to the most visible repair needs, not those presenting the most danger. Furthermore, Ceauşescu had verbally ordered that repairs of old buildings adhere to previous seismic stability regulations, not those recently established for new construction, and that engineers stop taking seismic stability measurements or including them in reinforcement plans. The Inspector General's Office reduced the amount of money previously allocated for reinforcement work.[24] A telex sent on July 11 to engineers at the Inspector General's Office for Construction reiterated that building assessments would not have new or additional seismic stability measures and added that if these were included, the assessments would be annulled: "Supplementary measures for earthquake safety of buildings cannot be introduced into reinforcement projects … all planned and current work will conform to this order … any assessments contrary to this will be canceled."[25] This telex reiterated that repairs and site cleanup had to be completed by August 23.[26]

Specialists in several domains faced serious challenges implementing the order. Its mandate raised tensions between engineers and other construction specialists and the bureaucrats charged with monitoring its implementation. Yet engineers and construction specialists also sparred among themselves. Some believed that the order conflicted with a new law for construction quality. It troubled experts, who believed that the superficial, minimal, and rushed repairs would leave people living in unsafe buildings and create a serious public risk in the capital (which it did!). They grappled with their moral responsibility to protect the public while working under an authoritarian leader who did not respect their scientific expertise and within a regime that had an extensive surveillance apparatus to deter undesirable behavior and encourage compliance. The order required specialists to ignore necessary structural repairs in favor of cosmetic ones. Some civil engineers, urban planners, and others were especially concerned about compliance with the order as individual building repair needs varied, yet the order's directives were general in scope.[27]

In the days, weeks, and months following the earthquake, engineers and construction specialists did the best they could with shifting official regulations and unofficial expectations. Civil engineers and other specialists completed what repairs they could with the time and materials available to them ahead of August 23. A supervisor at the National Research Institute for Construction told his repair teams to start the very next day by injecting epoxy into large cracks in support columns and "jacketing" them, meaning coating them with concrete of 8–10 centimeters. An engineer who completed repairs following the order, recalled "And we did! … in one day! Later on, we realized that among us there was [a Securitate informant] to report … we did the best that was possible for strengthening, under the limited conditions, of course."[28]

Some specialists were confused about how best to comply with the order, with many departments and individuals involved and not one ultimately responsible, including those at the eight city sector Executive People's Councils; the Central Institute for Research, Construction Projects, and Planning; the National Research Institute for Construction; and the State Inspector General for Investment and Construction.[29] Each was involved in the assessment and repair process and at times gave specialists contradictory information. For a centralized economy in the Romanian socialist form, the 1977 disaster response work was anything but centralized.

That summer, to meet the August 23 deadline imposed by the order, there was a ton of repair work to do and not enough time or people to do it. The Securitate noted, "the August 23 deadline for repairs means they no longer follow the reinforcement work plans provided, [and instead are] covering cracks with mortar without taking into account that some of them are resistance breaks."[30] This observation can be taken one of two ways. It may have been intended as confirmation that specialists were in compliance with the order, or it may have expressed concern that they were cutting corners and not doing their jobs correctly. Given the Securitate's structure and modus operandi, both interpretations are plausible; either the Securitate officers knew the order's directives or did not. Residents, too, were aware of the rushed and cosmetic fixes and joked about how specialists were using "anti-earthquake plaster," yet without ever being told exactly why.[31] While some specialists made only cosmetic repairs to hasten the work and comply with the order, most appear to have done as much as they were able within the circumstances; in both cases, however, specialists worried about their liability in case of the next strong earthquake.

Some specialists believed that the order violated a new construction quality law passed at practically the same time.[32] Only one week before Ceaușescu issued the order, Romania's Great National Assembly passed the construction quality law "for ensuring sustainability, operational safety, functionality, and quality of construction."[33] Specialists had to navigate between their liability under the new law and their responsibilities to comply with the order. The 1977 construction quality law applied to new residential and industrial construction, but specialists were particularly concerned about their liability under two of its articles when addressing the reinforcement needs of buildings damaged by the 1977 earthquake. Article 4 required that all "projects adhere to the seismic zone guidelines outlined in the law," referring to new guidelines assigned by geographic zones, and it obliged planners and builders to mention in their plans "the degree of seismic and fire resistance for which they were designed and built."[34] Article 14 had the potential to hold specialists liable, stipulating that "workers, foremen, technicians, and engineers in the design and execution departments must strictly adhere to the design and work plans and are obliged to make a rigorous control of their work and to remove any deficiency produced."[35] Would there be deficiencies in their work when complying with the order that might put them in violation of the new law? The construction quality law purportedly applied to new construction only, but how could specialists know whether it would be used to litigate their disaster recovery structural reinforcement repairs?

The July cessation order put engineering and construction specialists in a tough place. If they followed it, thousands of residents would continue to live in unsafe buildings, with a false sense of security from the cosmetic repairs. If they did not follow it, retribution or punishment was a real possibility. While the regime, and Ceaușescu in particular, appeared unconcerned about the public risk from the hastened, cosmetic, and incomplete repairs demanded by the order, many specialists were worried. In early August, as reported by the Securitate, four unnamed engineering specialists took a stance against the order, considering themselves "morally responsible for human life and society in the event of a new earthquake and … dissatisfied with nonspecialists making decisions at worksites while the responsibility was placed on civic professionals and structural engineers."[36] Shortly after receiving the order, one architect at the capital's urban planning institute said that "the [deadline] established to finish the reinforcement work in the capital is very short because of the sheer volume of work and the fact that they have not enlisted the specialists stipulated by law for documenting the damage." Yet, many specialists recognized not only that they were liable, but also that the order was wrong, as one architect shared about complying with it: "He who proceeds as such is committing a crime."[37] The order to end assessments, to rush to finish repairs, and to cover all visible damages within six weeks resulted in an unknowable number of unaddressed repairs.

Gheorghe Ursu's Exposure of the July Cessation Order

The civil engineer Gheorghe Ursu is the one known specialist to have publicly shared information and expressed concerns about the ramifications of the July cessation order. He made several efforts to alert Bucharesters and Romanians to the public risks posed by it, and his 1985 murder at the hands of the regime is understood to be connected to this public criticism. In 1979 and 1984, writing under pseudonyms, he sent letters describing the July 4 meeting and explaining the ramifications of its order to Radio Free Europe (RFE), which broadcast the information in its Romanian programming.

Ursu served as a lead engineer for a Bucharest building assessment commission. In 1977 he worked at the Bucharest Research and Design Institute for Systemization, Housing, and Municipal Engineering as a civil engineer and specialist in seismic engineering, as well as an urban planner and manager. A participant at the July 4 meeting, after the order's issuance, criticized the reduced reinforcement plans and structural assessments, questioning his own and his colleagues' possible liability under the new construction quality law. The Securitate documented his critique, specifically the mismatch between the reinforcement work recommended by his commission and the State Inspectorate's final approval of far fewer repairs.[38] It also tracked Ursu's structural assessment of one building, the Patria. Significantly damaged in the 1977 earthquake, the 1937 building sits on a corner, a factor that increases any building's seismic risk. The ground floor housed a single-screen movie theater and shops; its residents lived in its fifty-four apartments, located in the eight- or ten-story towers. During the hastened work under the order, the Securitate noted that Ursu and three colleagues said that the repairs authorities reduced were "contradictory and completely confused the planners." Authorities cut the budget by half, reduced the existing number of support pillars determined to need repair by 41 percent, and reduced the approved number of additional, new resistance columns by 88 percent and new support beams by 74 percent.[39] Almost a month after the deadline to complete repairs, the Securitate was still keeping tabs on Ursu and his team, which continued to work at the Patria despite the deadline having passed. The Securitate reported that Ursu "declared in front of his colleagues that he 'will appeal to the [State] Inspectorate for two more [resistance] columns even if he has to pay for them himself (around 6,000 lei).'"[40] This amount equated to roughly $300, a significant amount for one person to contribute, equal to three months of an average salary, well beyond Ursu's means.[41] Ursu ultimately did not sign off as the approving engineer of the Patria's reduced repair plan. While the Patria still stands today, it has been listed since the early 1990s by the Bucharest mayor's office as among the almost four hundred buildings expected to collapse in the next major earthquake. Its cinema has finally closed to the public, and structural reinforcement repairs have started.[42]

Ursu did not let the risk surrounding the insufficiently repaired buildings rest. Without his efforts, Romanians and Bucharesters may not have known about the public risk associated with the order. In 1979, on the second anniversary of the earthquake, RFE's *Povestea Vorbei* (The Tale of the Word) program featured the text of a letter from Ursu that described the July 4 meeting and the order's implementation and ramifications.[43] Ursu admitted to the Securitate that he wrote letters to RFE, and during a 1985 interrogation he said:

> I must do something to reinstate some of the reinforcement work.... I believe that by [my] presenting my viewpoint on RFE, some parts of our country will become alert and will take action.... I drafted some material with the conclusion that in Bucharest sufficient reinforcement to ensure the stability of buildings damaged in the event of a new earthquake had not been made.[44]

More than five years later, in October 1984, RFE broadcast a program about the public risks in Bucharest that remained from the July cessation order, using text from a letter signed by "a group of Romanian tourists in the Federal Republic of Germany."[45] Ursu also wrote that letter, in which he described the reduced repairs as well as the unpaid overtime and Sunday shifts demanded of many workers in the name of the 1977 earthquake disaster response. Four days after the broadcast, Securitate officer Vlad Iulian circulated the department's transcription of it, along with a memo that asked, "How were the letters sent abroad? What do you think needs to be done in this case?"[46]

In December 1984, two months after RFE broadcast information from Ursu's second letter, the Securitate searched his office and home and confiscated notebooks, journals, letters, poems, travel notes, and movie and theater playbills. Over the first nine months of 1985, the Securitate interrogated Ursu as many as twenty times. Then, in September, the police arrested him on the premise that he possessed seventeen dollars; it was illegal for most Romanians to have foreign currency. They held him in the shared offices of the Bucharest police and Securitate, where officers interrogated him and instructed his cellmate to beat him. To treat injuries from the beatings, so severe they required surgery, the regime moved him to Jilava Prison near Bucharest. On November 17, 1985, Ursu died there from abdominal trauma: he had been beaten to death.[47]

Gheorghe Ursu's case is important to understanding the Ceauşescu regime's 1977 earthquake disaster response as he was the lone voice that revealed to the public the risks associated with the regime's July cessation order and the insufficient repairs to the many seismically at-risk buildings. He told the Securitate that the letters broadcast by RFE expressed his own "feelings to take a stand against the situation, considering it my conscious duty to express these feelings."[48] He felt he had a moral responsibility to speak up, and he believed that someone who heard it might also recognize the risks to Bucharesters and do something about them. That did not happen until after the end of the Ceauşescu regime.

Challenged Implementation of the July Cessation Order

Implementing the July cessation order was confusing for specialists, regime officials, and Bucharest residents alike. Members of the assessment commissions were unsure about reporting procedures, their liability under the new construction quality law, the new parameters to determine structural safety, and the repair plans to follow. To comply with the order, agency-level officials reduced or eliminated specialists' initial reinforcement recommendations. In one case similar to the Patria, officials drastically reduced the repair plans: reinforcement of only four structural support columns instead of twelve; "jacketing" reinforcement of three pillars rather than thirty-seven; and repair of five resistance columns instead of the thirty-seven initially recommended.[49] During those frantic weeks before the August 23 deadline, specialists fought for required repairs and requested reinstatement of the original assessment recommendation plans or, at least, additional repair work beyond what was finally approved. While some cases of additional work being approved did happen, this was not the norm. In one case, the state inspector general returned to a worksite and approved additional "jacketing" of several columns in one building.[50] In another, authorities doubled the funds to pay for a building's reinforcement work because the director of the Bucharest Industrial Construction Center requested it. In this case, construction specialists had appealed for approval to repair three additional resistance columns beyond the four approved, even though, as the Securitate noted, "the specialists consider that even this would not sufficiently solve the problem for this building."[51] In short, that summer, administrators rarely approved repair plans at their pre-July assessment levels.

Residents of damaged buildings did not know the specifics of the order, and many were confused, frustrated, or angry by the process that summer. Assessment commissions no longer came. Workers arrived with original plans changed. Bucharest authorities approved fewer repairs. And the revised repair plans were not shared with residents, which often "created the possibilities for unhappy and unfavorable commentaries."[52] One resident said that her city sector council "was playing with people because after members told people to pack their bags, then they told them to stay" and wait for another commission to evaluate their building again.[53] Residents of a damaged building were upset that one commission told them not to make other repairs until structural reinforcement was done, then other officials canceled all reinforcement work, and finally experts told them to proceed only with repairs. Residents were upset not only because of the change in plans but also because "new cracks appeared and old ones widened," increasing their fear about their building's structural integrity.[54]

Ceauşescu was aware that some specialists complained about having to implement the work demanded in the order. One case is particularly illustrative. Specialists were concerned that repairs approved for the building at 36 Maria Rosetti Street did not meet the "minimum" proposed by the city's urban planners, a situation, noted by the Securitate, "that placed more than one hundred residents at risk and has the potential for a future serious accident."[55] The Securitate's report described the initial repair plan and the insufficient repairs completed, and it went on to recommend that repairs continue as in the original plan. Ceauşescu did not like what he read. Ten days after the Securitate drafted the document, someone handwrote across its heading: "The first secretary did not appreciate this note and ordered that we report to him information about the planner, the source, and if he is an officer of the Interior Ministry." The report's annotator instructed the writer to "come to me now!" to discuss Ceauşescu's lack of "appreciation" and the source, whether an informant or Securitate officer.[56] As of 2024, the building at 36 Maria Rosetti Street, an impressive, nine-story apartment building with thirty-one units, situated at the end of a block and built in 1936, was still on the list of structurally unstable buildings, unchanged from its 1994 assessment.[57]

Conclusion

The July cessation order required regime authorities and specialists working on the buildings damaged by the 1977 earthquake to end structural assessments, reduce approved reinforcement plans, and hasten repairs already underway, prioritizing the most visible damage, all by August 23. The order presented construction and engineering specialists with a dual dilemma: they could be held liable for insufficient repairs under a new construction quality law passed at practically the same time, and they had a moral obligation to repair buildings they knew were unsafe. They navigated the order's directives, which regime bureaucrats orchestrated and approved, a process that highlighted divisions between regime bureaucrats and construction specialists. Bureaucrats reduced assessment plans while some specialists requested reinstatements. The regime's multilayered and redundant bureaucracy, especially given the city's administrative structure across eight sectors, made it difficult for specialists to know which department or manager to appeal to.

As if it were using the 1940 post-earthquake experience as a template, the regime was concerned about reinforcement at first but then dialed down its efforts because Ceauşescu thought there were too many evacuations, repair teams, and assessment commissions fanned across the city. Rather than ignore experts' findings, as the fascist Antonescu government once did, the Ceauşescu regime first reduced the repair guidelines and assessment numbers and then finally reduced approved interventions, hastened repairs, and stopped assessments of damaged buildings altogether. It swung the pendulum of risk assessments from one extreme to the other, where at first teams overestimated the number of buildings to be evacuated only to finally approve and complete far fewer than needed, abruptly ending the 1977 earthquake disaster response. It did so because Nicolae Ceauşescu was upset by what he perceived as "chaos" and "panic" from the disaster response efforts that spring, all to meet his demand for the city to be cleared in time to celebrate the August 23 holiday. He chose cosmetic fixes, ignored the public risk the damaged buildings posed, and dismissed construction and engineering experts' recommendations. This decision has serious implications to this day. Authorities have identified almost one thousand residential and mixed-use buildings in Bucharest at risk of significant damage in the next major earthquake, with a third expected to collapse. The information bravely shared with the public by the engineer Gheorghe Ursu about the July cessation order, which RFE broadcast, alerted Romanians to its ramifications:

> If March 4, 1977, remains in the Romanian consciousness as a date evoking cataclysm ... the date of July 4, 1977, means something only to us insiders. For a small number of people—structural engineers and others responsible at different levels for relieving the earthquake damage—July 4, 1977, means a sort of second earthquake.[58]

The public risk in the case of Romania's next major earthquake is directly linked to the order that ended the Ceauşescu regime's disaster response, one exacerbated by post-1989 governments' start-and-stop initiatives to support structural repairs (Chapter 7).

Notes

1 Secţia Economică, 78/1977, 1–31, FCCPCR, ANR.
2 Secţia Economică, 78/1977, 30.
3 Administraţia Municipală, "Listă imobile expertizate."

4 *Povestea Vorbei*, no. 195, 3–4, RFE/RL Romanian Broadcasting Department records, box 3811, folder 2, HILA.
5 Secţia Economică, 78/1977, 10–10 rev., FCCPCR, ANR.
6 Secţia Economică, 78/1977, 11 rev.
7 Secţia Economică, 78/1977, 4.
8 Hossu Longin, *Memorialul durerii*, 358; and Tismăneanu, *Stalinism for All Seasons*, 205.
9 Secţia Economică, 78/1977, 6, FCCPCR, ANR; *Povestea Vorbei*, no. 195, 3.
10 Secţia Economică, 78/1977, 4, FCCPCR, ANR.
11 *Povestea Vorbei*, no. 195, 6, RFE/RL Romanian Broadcasting Department records, box 3811, folder 2, HILA.
12 Secţia Economică, 78/1977, 25, FCCPCR, ANR; *Povestea Vorbei*, no. 195, 5.
13 Secţia Economică, 78/1977, 29, FCCPCR, ANR.
14 Secţia Economică, 78/1977, 25.
15 Secţia Economică, 78/1977, 26.
16 The Illinois Academy of Sciences did not offer Elena Ceauşescu honorary membership in 1978, as is often cited. Robyn L. Myers, email message to author, September 20, 2016.
17 Secţia Economică, 78/1977, 28, FCCPCR, ANR.
18 *Povestea Vorbei*, no. 195, 5, RFE/RL Romanian Broadcasting Department records, box 3811, folder 2, HILA.
19 Secţia Economică, 78/1977, 29, FCCPCR, ANR.
20 Secţia Economică, 78/1977, 26.
21 Secţia Economică, 78/1977, 33.
22 D.13.339, v. 37, 381, DB, CNSAS.
23 D.13.339, v. 37, 370 and 381.
24 D.13.339, v. 37, 387–89 reverse.
25 D.13.339, v. 37, 381.
26 D.13.339, v. 37, 371 rev. and 381, DB, CNSAS; Secţia Economică, 78/1977, 36, FCCPCR, ANR.
27 D.13.339, v. 37, 387, DB, CNSAS.
28 Emil-Sever Georgescu, email message to author, May 18, 2016.
29 D.13.339, v. 37, 371, DB, CNSAS.
30 D.13.339, v. 37, 371.
31 *Povestea Vorbei*, no. 195, 4, RFE/RL Romanian Broadcasting Department records, box 3811, folder 2, HILA.
32 D.13.339, v. 37, 370 rev. and 389 rev., DB, CNSAS.
33 E.-S. Georgescu, "Earthquake Engineering Development," 8; Lege privind asigurarea durabilităţii, sigurantei în exploatare, functionalităţii si calităţii construcţiilor, law no. 8, July 1, 1977.
34 Lege privind asigurarea durabilităţii, article 4.
35 Lege privind asigurarea durabilităţii, article 14.
36 D.13.339, v. 37, 372–73 rev., DB, CNSAS.
37 D.13.339, v. 37, 381 rev.
38 D.13.339, v. 37, 373.
39 D.13.339, v. 37, 373.
40 D.13.339, v. 37, 389 rev.
41 Andrei Ursu, pers. comm., November 20, 2016.
42 Administraţia Municipală, "Listă imobile expertizate."
43 *Povestea Vorbei*, no. 195, 1–9, RFE/RL Romanian Broadcasting Department records, box 3811, folder 2, HILA.
44 Eskenasy, "Cazul Gheorghe Ursu."
45 D.21, v. 55, 110–13 and 120, DB, CNSAS.

46 D.21, v. 55, 120.
47 Deletant, *Ceauşescu and the Securitate*, 331.
48 Eskenasy, "Cazul Gheorghe Ursu."
49 D.13.339, v. 37, 389–89 rev., DB, CNSAS.
50 D.16.334, v. 5, 5 rev., DB, CNSAS.
51 D.16.334, v. 5, 7–8.
52 D.13.339, v. 37, 103, DB, CNSAS.
53 D.12.639, v. 15, 86 rev., DB, CNSAS; D.13.339, v. 37, 103 rev., DB, CNSAS.
54 D.13.339, v. 37, 381 rev., DB, CNSAS.
55 D.13.339, v. 37, 374.
56 D.13.339, v. 37, 374, and see 374–75 rev.
57 Administraţia Municipală, "Listă imobile expertizate."
58 *Povestea Vorbei*, no. 195, 1, RFE/RL Romanian Broadcasting Department records, box 3811, folder 2, HILA.

7 Living Unsafe

Bucharest's Public Risk

Today a serious public safety risk exists in Bucharest, Romania. Almost four hundred buildings damaged by the 1977 earthquake remain structurally unsound, expected to collapse in the next major earthquake.[1] Additionally, about one thousand are expected to experience significant damage and another 1,500 some damage. That public risk is a direct legacy of the Ceauşescu regime's 1977 earthquake disaster response, and specifically its July cessation order, which ended seismic structural assessments and hastened repairs to the structurally unsafe buildings. We do not know when the next major earthquake, one registering 7.0 or stronger on the Richter scale, will occur, just that it will. Damage to built structures, contributing to injury and death from the full or partial collapse of potentially hundreds of residential and commercial buildings, is inevitable as Romania is one of the most seismically active places in Europe.

Since the end of the Ceauşescu regime in December 1989, Romanian national and municipal governments and scientific experts have addressed that risk with a range of mitigation efforts. They have drafted new laws and city ordinances and created public awareness campaigns. They have strengthened building codes for earthquake safety; created a graduated, tiered seismic risk classification system and assigned damaged buildings to it; and provided grants and low-interest loans, technical assistance, and other supports for structural repairs. Romanian scientists have world-class data detection mechanisms to inform sophisticated public warning systems. They run information campaigns to educate the public about the risks, actions to alleviate them, and ways to prepare for the next major earthquake. Together, these efforts will help reduce casualties from structural damage when the next earthquake shakes the country, yet many in Romania do not know about or are unphased by the risk.

Making It Personal

In July 2014, a knock at my door brought the reality of an earthquake's destructive potential home. That summer I sublet an apartment in Bucharest to work in Romania's political police, national, and local archives. At the door was one of my neighbors, flustered and exuding a sense of urgency. My small

DOI: 10.4324/9781032632933-8

children knew her as the owner of Fetiţa, the three-legged dog they liked to pet. We exchanged greetings, and then she immediately pivoted to her reason for visiting. "You are here with small children, I felt I had to share this with you," she said, insisting that she show me something in the basement. Not to scare them, I told my children that Fetiţa's mom wanted to show us something, leaving out the danger in her tone.

Our small apartment was one among thirty in a striking multistory art deco building from 1936. Like many buildings of its era, it had a rundown exterior, neglected interior common spaces, and an elevator so small my children and I barely fit. The apartment had a tiny open living/kitchen area, a bathroom with an excellent washing machine, and a small bedroom with windows that looked out on the long balcony with a terrifyingly low wall, where my children liked to frighten me by getting too close to its short railing. We hung our laundry there, watched the sunset, and played "drive-through" using the bedroom window. The building's location was ideal. Situated in the city center, it was steps away from one of the archives where I researched, the city's central park where we went every evening, a well-stocked small grocery, public transportation, and a pretty good *covrigarie*, or soft pretzel stand.

My kiddos and I followed our neighbor, took the elevator down to the ground floor, walked down the stairs to the basement, and proceeded directly to one of the building's ten massive support columns. It must have been at least three feet in diameter. She walked toward it, literally reached inside, and pulled on its iron rebar support cables. As she held and twisted the rebar, concrete crumbled to the floor. "You and your children must move immediately; the city inspectors came and reclassified us into seismic risk Class I because four of the ten support columns need retrofitting." I was in Bucharest to research the Ceauşescu regime's 1977 earthquake disaster response and knew that a building assessed into seismic risk Class I, without prescribed repairs, was condemned to collapse in the next, inevitable, major earthquake.

That night I could not sleep. I knew what a strong earthquake could do to apartment buildings like ours. Earthquakes are frequent in Romania. Since the devastating major ones in 1940 and 1977, Romanians have experienced half a dozen moderate-strength (5.0–6.9 on the Richter scale) earthquakes—in 1986, 1990, 1991, 2004, 2014, and 2016—contributing to a few deaths, a few hundred injuries, and minimal damage to built structures. I had seen devastating images of the collapsed and damaged buildings, rescue teams, general chaos, and destruction from both the 1940 and 1977 earthquakes.[2] I had read and heard harrowing firsthand stories from survivors who were trapped in buildings, unsure how to respond during the tremor, and who lost loved ones. Before renting the apartment, I had researched the building and learned it had been assessed into seismic Class II, likely to experience damage but not to collapse. What happened that downgraded it to Class I? Regardless of why, I trusted the new assessment and the crumbling concrete I saw. My mind went immediately to worst-case scenarios: our building collapsing, crushing us in our sleep. Or my children surviving without me, injured, wandering on their own, unable to communicate in Romanian, shocked and lost. Or, my survival without them. I quickly prepared a "bugout bag"—a backpack with food, water, cash, clean underwear, a phone charger, and our passports—moved it next to the bed I would then share with my children, and crawled in between them. If an earthquake came, we would be together.

Seismic Risk Factors

A most tragic and important legacy of the 1977 earthquake and the Ceauşescu regime's disaster recovery is the extreme danger posed by the thousands of seismically structurally unsafe buildings in Romania's capital of Bucharest. Home to 2.1 million people, Bucharest has a high population density, just over 7,000 people per square kilometer.[3] The primary danger from an earthquake is how much the earth's movement affects built structures, causing their full or partial collapse and resulting in injury or death to those inside or in their way. The extent of this risk results from a combination of seismic, engineering, construction, political, and social factors.

The most obvious factor is seismic: the stronger the earthquake the more potential damage. Seismologists register an earthquake's strength on the Richter scale, which categorizes an earthquake at 3–4.9 as minor or light, 5–6.9 as moderate, 7–7.9 as major, and 8 and higher as strong, with each integer on the scale representing a ten-times increase in strength from the one before. Another factor is how deep underground the earthquake causes the earth to move. Engineering and construction risk factors concern the built structures an earthquake might affect. A built structure's seismic vulnerability is calculated by considering all the earthquake-specific factors to predict how likely it is to experience damage from an earthquake; this might include the type of construction materials and quality used and its degree of compliance with building code that considers seismic resistance. Many deaths after the 2023 earthquakes in Morocco and Afghanistan were the result of the collapse of structures made with earthen brick, which crumbled into dirt and dust, leaving few to no air pockets for possible survival when buried. In Romania's 1977 earthquake, when buildings made of concrete slabs collapsed, the floors fell one on top of the other, or "pancaked," creating some air pockets, where some survived until rescued. A building's seismic resistance is also affected by modifications made, like removing (load- or nonload-bearing) walls, cutting new windows, or creating more usable square footage by removing ground-floor support pillars. One modification that contributed to the collapse of some buildings damaged by the 1940 and 1977 earthquakes was the removal of load-bearing columns on the ground floor to create more usable square footage for retail shops or cafés. The building's age, too, can be a factor, not because time inevitably affects resistance, but because the era when it was built may be known to have had construction quality deficiencies or building codes commonly skirted. This factor contributed to the majority of the buildings that partially or fully collapsed following the 1977 earthquake, due to lax construction practices and building code compliance of the historical era when they were built. Many of the more than fifty thousand deaths in Turkey from the February 2023 earthquake were avoidable as experts had been warning of the risks from unsafe buildings for decades.[4]

In addition, social factors also affect the potential for death and injury from earthquakes. Romanian scientists stress that "socio-economic losses could be far greater if a single ten-story (apartment building) collapses (which … happened during the 1977 earthquake), compared to the collapse of 50 small houses."[5] When there are more people in and around vulnerable buildings, the risk increases. For that reason, an area's population size and density also contribute. The likelihood of death and severity of injury is further affected by the ability of first responders to reach victims. Traffic congestion and difficult-to-access areas can hamper first responders' ability to reach damaged buildings. A reduced, ill-equipped, or poorly trained first responder force, inadequate or underused alert systems, and lackluster public information campaigns, too, increase the risk of injury and death. The time of day and season of the year can increase risk as people are more likely to be inside at night and during colder months. People in marginalized communities with access to fewer resources to mitigate and prepare for risk, and those who lack awareness or are in a state of denial of potential earthquake damage (and are therefore most unprepared) are also at increased risk. The public's awareness, understanding, and acceptance of the potential risk from earthquakes and their ability to prepare are key (Appendix C). Lastly, for earthquakes, their extremely short warning time—in Romania only twenty to twenty-five seconds at the maximum—also contributes to one's risk of injury or death.

I knew all this before renting a place to live in Bucharest that summer. I had stayed in the same building during an earlier research trip. All of this I arranged with the help of a friend, whom I knew from when I had been a Peace Corps volunteer in neighboring Moldova. For an earlier trip, he helped me stay with his neighbor, a hospitable, kind, elderly widow, who had lived in a spacious two-bedroom apartment her entire adult life. My friend had assured me that the building was safe, having been assessed into seismic risk Class II, and it was listed as such in the mayoral office's online list of at-risk buildings. He believed that buildings constructed before 1940 that had survived both the 1940 and 1977 earthquakes, like his, were especially resistant. For the summer in Bucharest with my children, I sublet his apartment. He was not trying to sell me on the sublet, just the fact that, in his opinion, it would be okay. I accepted his argument. Then my neighbor showed me the state of the support columns in the basement, reclassifying it as Class I.

Postcommunist Risk Mitigation

Because the Ceauşescu regime ended its disaster recovery efforts after only four months with its July cessation order, little was done during the end of the communist era to mitigate the potential damage, injury, and death from the next major earthquake. After 1989, however, Romania's postcommunist national, municipal, and local governments made significant efforts to reduce the public's risk from the hundreds of seismically unsafe buildings. They passed new legislation and amended existing building codes, assessed buildings damaged by earthquakes, assisted building owners in drafting plans and beginning repairs, and created a tiered seismic risk classification system. Romania has more than four dozen laws concerning seismic and other disaster risks' safety, mitigation, response, and warning. In the decades since communism, experts have updated and increased the number of seismic detection devices and developed warning and alert systems. Working in collaboration, nongovernmental organizations, various levels of government, and scientists have also advocated for and organized public information campaigns.

While the Ceauşescu regime did not repair most of the buildings damaged by the 1977 earthquake, it implemented some policies to mitigate future risks. In 1978 and 1981 it revised building codes for new construction in the capital, requiring stability in the case of an earthquake measuring at least 7.0 on the Richter scale. Postcommunist governments, however, established even more stringent and specific seismic resistance building codes in 1991 and 1992, revising these in 1997. They increased the resistance requirements to at least 7.0 for new buildings constructed in communities outside of Bucharest and west of the Vrancea Seismic Zone, Romania's seismic epicenter; new buildings in Bucharest, depending on the neighborhood, were required to be able to resist earthquakes measuring 8.0 or higher on the Richter scale.[6]

Today, Romanian governments, scientists, academics, and nongovernmental organizations continue the work of risk mitigation from education and information to preparation and support for repair. The country has several government agencies dedicated to earthquake and disaster detection, mitigation, and information sharing. The Ministry for Development, Public Works, and Administration (Ministerul Dezvoltării, Lucrărilor Publice şi Administraţiei, MDLPA) has the primary role in mitigating and managing earthquake risk, including investigations, assessments, and response plans. Its State Construction Inspectorate (Inspectorate de Stat în Construcţii, ISC) oversees urban planning, construction authorization, and compliance with building codes. The National Research and Development Institute for Geophysics (Institutul Naţional de Cercetare-Dezvoltare pentru Fizica Pământului, INCDFP) monitors seismic activity through its National Seismic Network and the National Data Center and holds a primary role in the development and maintenance of the earthquake warning systems. The National Institute for Research and Development in Construction, Urbanism, and Sustainable Territorial Development (Institutul Naţional de Cercetare-Dezvoltare în Construcţii, Urbanism şi Dezvoltare Teritorială Durabilă, URBAN-INCERC) researches and develops activities for the public about construction, urban planning, sustainable territorial development, and housing, and the National Seismic Monitoring Protection Network for Historical Structures (Reţeaua Naţională de Monitorizare şi Protecţie Seismică a Patrimoniului Construit), provides notices, warnings, and alerts.

The Seismic Risk Classification System

In May 1990, just months after the end of the Ceauşescu regime, the postcommunist government began to assess and inventory the seismically unsafe buildings as a result of the earthquakes in 1940 and 1977. Initially, it assigned buildings into one of three "urgency" levels—U1, U2, or U3—indicating the estimated time needed to complete structural stability repairs, a classification system also used by the Ceauşescu regime in its 1977 earthquake disaster response. In January 1994, recognizing the need for a system that reflected seismic structural risk as opposed to temporal exigency, municipal authorities required that damaged buildings be evaluated for seismic structural stability and classified into a four-tier seismic risk classification system, defined as:

Class I: those likely to fully collapse in case of an earthquake
Class II: those highly likely to experience severe structural damage, endangering residents, but total or partial collapse unlikely
Class III: those likely to experience moderate damage that may endanger residents
Class IV: those that would experience some damage[7]

The system does not define the minimum strength of the earthquake for full or partial collapse or damage, trusting that determination to the experts who conduct the assessment. With the new seismic risk classes established, experts went out again and assessed hundreds of buildings for seismic stability. In Bucharest, starting in 1994, experts first assessed buildings five stories or taller that had been constructed before 1940. In September 1997, the government required the Bucharest mayoral office to organize and publish a list of the buildings assessed and assigned to the new seismic risk classes and the earlier urgency assessment levels.[8] The office's Municipal Administration for Retrofitting Buildings at Seismic Risk (Administraţia Municipală pentru Consolidarea Clădirilor cu Risc Seismic, AMCCRS) hosts, frequently updates, and makes the list available to the public.[9] Basic data includes the assessed building's address, neighborhood (sector), year built, number of floors and apartments, and notes about whether it has a basement, mezzanine, or workshop/shed. Seismic stability data includes the year or years of technical assessments, names of engineers who did this work, seismic risk class assigned, and observational notes, which might include any previous assessment information and whether its seismic risk class assessment has changed. The very few buildings that were once assessed into a seismic class or urgency level and have since been retrofitted are listed as "consolidated"—that is, as having completed the necessary structural stability repairs. As of April 2024, the list includes 2,758 buildings, with only 111 marked as consolidated and 5 partially consolidated. Five percent (142) of those on the list have been assessed but not assigned a seismic class or urgency level. Of the 2,500 that have been assessed and not yet been retrofitted, 60 percent (1,497) are assigned an urgency level and 40 percent (1,003) a seismic risk class. About 40 percent (395) of this latter group are in Class I, those likely to fully collapse in the next earthquake.[10]

Twenty-nine, just 1 percent, of the buildings have been reassigned from an urgency level into a seismic risk class. Reassessment has not been done, mostly, because the government prioritized assessing buildings of five or more stories built before 1940, and most buildings classified into an urgency level have just one story. That does not mean that they are not at risk of significant damage or collapse from the next earthquake. We can only assume that their collapse would affect fewer people.

Legislation passed in 2022 continued the government's commitment to seismic risk mitigation.[11] It requires private and public owners of buildings at potential seismic risk (whether individuals or groups) to have their buildings technically assessed by engineers. If buildings do not meet seismic resistance requirements, they are classified into a seismic risk class or urgency level and owners are required to begin the repairs prescribed to meet the current code. Building owners of those assessed into any one of the four seismic classes are required to promptly inform local authorities and the National Agency for Taxation Registry and Real Estate Publication (Agenţia Naţională de Cadastru şi Publicitate Imobiliară, ANCPI) of the new designation; owners of those assessed into Class I have sixty days to do so. Once a building is assessed into Class I, owners have two years to prepare a repair plan and begin the seismic stability work outlined in it, meeting the minimum seismic resistance level as outlined by law at the time the repair plan was approved.

Condemned to Collapse: The "Red Dot" Buildings

When walking through Bucharest's city center one might see giant red circular signs affixed to buildings' exteriors (see Figure 7.1). These "red dots" (*bulinele roşii*), as they are colloquially known, mark buildings assessed into seismic Class I, anticipated to collapse in the next major earthquake. The giant, circular signs, sixty centimeters in diameter, appear above building entryways, three or four meters above the ground, and read: "Building technically assessed as seismic risk Class I" (*clădire expertizata tehnic încadrata în Clasa I de risc seismic*). Starting in 1996 in Bucharest and 2007 all across Romania, as one way to alert people of the risk they present, buildings assessed into Class I must display these "red dots," yet very few do.[12]

The idea for marking the most at-risk buildings came from Nicolae Noica, a government minister who in June 1996 was named to the newly created Central Commission for Prevention and Defense against Seismic Effects and Landslides (Comisiei Centrale pentru Prevenirea şi Apărarea Împotriva Efectelor Seismice şi Alunecarilor de Teren).[13] One of the commission's responsibilities was to "ensure the public is informed through mass media about the potential risk, the imminent occurrence of earthquakes or landslides, their effects, as well as preventative measures taken … to propose mechanisms and tools for risk mitigation, as well as methods for identifying and evaluating risks

Figure 7.1 The entryway of 38 Brezoianu Street, Bucharest, with a "red dot" indicating its Class I seismic risk assessment, completed in 1995. Photo by author, June 26, 2018.

in the field of earthquakes and landslides."[14] One mechanism it employed was the "red dot." The initial objective of marking buildings was to raise awareness of the risks they posed to owners, residents, neighbors, and the general public. The thought was that such a "scarlet sign" might encourage action to make the necessary structural repairs. Several unintended consequences, however, have arisen: heightened fear, decreased real estate values, fraud, deception, and corruption to avoid having a "red dot" affixed to one's building.

At their best, these signs inform residents, passersby, potential owners, and renters of the significant risk from buildings around them in case of an earthquake. At their worst, they shame building owners. Owners can experience decreased property values, lower rents, and lost business revenue when a "red dot" marks their building. It does not necessarily fast-track the structure for repairs. Over the years rent in "red dot" buildings has been significantly lower than in neighboring unmarked buildings. Some Bucharesters have seized this opportunity to live in the city center, where many of the seismic Class I, "red dot"–marked buildings are located. Yet many residents have no choice if they do not have the means, ability, or motivation to move to a safer building. As of 2024, almost four hundred buildings remain in seismic risk Class I, eligible for the "red dot," yet few display one. As of February 2024, the building I moved out of after seeing the concrete crumble from its support column does not display one as it is still listed in seismic risk Class II from its first assessment, in 2000. The downgrade my neighbor alerted me about has not been publicly reported or recorded.

When I learned that our building was downgraded to seismic risk Class I, I spent all my waking moments searching for a new place. I looked at apartments available for short-term rental in a convenient location and cross-checked their addresses with the official list of buildings in the urgent and seismic risk classification system. No luck. All were on the list.[15] I then searched home-share sites and found a conveniently located apartment on the north side of the city's central park, in a nine-story building with twenty-six apartments built in 1936. Because of its age and height, I was not hopeful when I checked the list. It had been assessed, yet its seismic class was not listed because it had been structurally reinforced in 2014. I had found an apartment in a safe building! But did I? Could I trust the information? One hears stories about buildings placed in a safer seismic class than their true assessment (like our first building), with the required structural stability work incomplete, nonexistent, or not assessed by certified engineers. As for buildings required to display a "red dot," owners can and do bribe officials so that it is never mounted. Or, owners remove the "red dot" themselves. I saw one "red dot" painted over in a similar color to the building's exterior, a poor camouflage attempt (see Figure 7.2). Since I photographed it in 2014, that painted-over "red dot" has been removed, yet the ten-story building, situated just steps away from the central park and built in 1948, remains in Class I seismic risk, recently reassessed in 2022, twenty-nine years after its first assessment.

Verification of retrofitting is vital as corruption among public officials is still common in Romania. In 2023, Transparency International found that 20 percent of Romanian households reported having paid a bribe to access public services and 45 percent thought that corruption had increased in the previous year.[16] Romania scored 46 on Transparency International's Corruption Perception Index, a measure of how corrupt experts and business people perceive their country's public sector, with a 0 score as "highly corrupt" and 100 as "very clean."

As for the building I found on the home-share site, in 2014 it was listed alongside seventy-nine other buildings in Bucharest that had been assessed as unsafe, but repaired and brought to seismic stability code. In the ten years since, thirty-two more buildings from the list have been retrofitted. To verify that the building I found had been retrofitted, I contacted a friend's mother who lived on the same block. She confirmed that the reinforcement work had happened and remembered that residents had moved out of the building for two years while it was done, recalling disruption to the neighborhood with extra traffic and obstructed sidewalks. I immediately messaged the host, secured the rental, explained to my friend why we had to leave the sublet, and moved the next day.

Government Assistance

In the initial years after 1989, in addition to conducting risk assessment, the Romanian government also helped owners of at-risk buildings assess and

Figure 7.2 A painted-over “red dot” above a handmade sign warning “Attention, falling plaster” at the entryway of the 1948 building at 5 Şipotul Fântânilor, Bucharest. This painted-over “red dot” has since been removed, yet the building remains in seismic risk Class 1, recently assessed in 2022. Photo by author, July 20, 2014.

retrofit their buildings, required if they were assessed into seismic risk Class I. Owners had access to twenty-five-year interest-free loans to pay for reinforcement work, as well as smaller benefits, such as a tax exemption for the building permit. For owners with a monthly salary lower than the state-calculated average, the state allowed a tax exemption for the loan payment as well.

The National Program for the Consolidation of Buildings with High Seismic Risk (Programul Naţional de Consolidare a Clădirilor cu Risc Seismic Ridicat, PNCCRS) administers several programs to assist individuals, businesses, local government agencies, and nongovernmental organizations with retrofitting their seismically at-risk buildings, including providing financial assistance for the technical assessment, repair, and upgrade of utilities. In 2023, PNCCRS allocated more than 600 million lei ($137 million) to retrofit twenty-four buildings.[17] Owners of buildings classified in one of the four seismic risk classes or three urgency levels can still apply for assistance from the mayoral office's AMCCRS. In 2021, AMCCRS responded to 1,235 requests, made 50 assessments, and ended the year with 180 open files. The programs are there, funded, and known, but the work moves slowly.

Even with the available government funding to assist, only a handful of owners of buildings assigned to a seismic class or urgency level have repaired their buildings. As of February 2024, only 4 percent of those buildings assessed into a seismic class or urgency level have been repaired for seismic stability. Bringing a multistory high-rise building up to seismic code is an arduous and expensive project. While the national and municipal governments have initiatives to assess and assist with retrofitting, the logistics and cost to do so have been prohibitive for most. As for the building we left after seeing its concrete crumble from a support column, the estimated cost of the recommended repairs was about $6,000, or $200 a unit. This was extremely low in comparison to the average for retrofitting repairs, which can range widely, anywhere from $100 to as much as $485 per square meter. Because many of the buildings are more than seventy years old, owners often upgrade heating, cooling, security, and plumbing systems as well, significantly increasing the repair costs. Experts agree on an average cost of about $200 per square meter. A typical one-bedroom apartment in Bucharest is 50 square meters, for a base cost for structural repairs of about $10,000, exponentially higher than the $200 asked of each unit in the building I left that summer.

In a 2016 case, city officials planned to bring to code an eight-story, twenty-five-unit apartment building constructed in 1915. The state offered to pay a private company about $8,400 per apartment to reinforce the building. The project moved forward without public comment or an open bid. The building was one of eighty included in a February 2016 "action plan" passed by Romania's Parliament, which committed the equivalent of $6.3 million to support the repairs. It should be noted that the government had drafted a similar action plan in 2015, and more than 80 percent of the buildings on the 2016 plan were carried over from that plan. The work moves slowly. Even more sobering, the buildings planned for reinforcement work in the 2016 action plan represent only 16 percent of the buildings classified into seismic risk Class I. Very few of the most at-risk buildings are in line for repair. One sticking point is the government's inability to exert pressure on private owners of buildings. In contrast, officials have authorized and repaired public buildings with some success. By 2016, structural reinforcement had been completed in the capital at about a hundred schools, a theater, the previous mayor's office, and an information center, as well as seventy-five public buildings outside the capital.

In addition to cost and systemic corruption, other challenges hinder the retrofitting of seismically at-risk buildings. There have been delays when buildings that require repair also hold historical significance or reside in areas protected for historical reasons. The ARO building, formerly the Patria building, sits at the corner of one of Bucharest's city center's largest streets. Since the mid-1990s, it has been assessed to seismic risk Class I and listed on the historical monuments and national treasures list because of its importance to Romania's architectural history. Finished in 1931, it is considered by architectural scholars and Romanian authorities to be the country's first important modernist building, designed by Horia Creangă (grandson of one of Romania's most noted authors), along with his wife and another colleague. Only in 2015 did legislation prohibit "large groups" from gathering in buildings classified as seismic risk Class I, closing the Patria's movie theater.

A range of other factors hamper repairs as well. Bucharest's population density contributes to slower and more costly retrofitting repairs because it is difficult to organize and carry out large-scale repair work in such populated and physically dense neighborhoods. Some owners ask for repairs when experts recommend demolition. There are not always enough workers certified for the assessment work. Some buildings are just too small or damaged that owners choose to do nothing rather than repair them, while other projects are stalled because of bureaucratic snags such as owners not submitting the correct paperwork or making mistakes on applications.

Legislation amended in January 2024 addresses some long-standing problems, with funding to subsidize rent and storage of personal items while a building is retrofitted.[18] It corrects a problem that has existed for decades: the inconvenience to residents when buildings are retrofitted. The work can take years. Residents have to move out and pay for a new place to live. Landlords lose rental income. Unanimous agreement among unit owners, required for work to move forward, can be difficult to attain and has stalled some requests for years. Since the most at-risk buildings are apartment buildings with many units, either with one building owner or a condominium organization with unit owners, this is a major obstacle. Owners don't see the value in it, so choose not to complete repairs. Some think the repair costs are not a good investment and prefer to buy a new and up-to-code apartment as the average cost is less or equal to reinforcement costs.

The biggest challenge of all, however, might be the denial that an earthquake, or damage from it, will, or can, happen. Some who own or live in seismically unsafe buildings feel repairs are unnecessary. They believe that they will not live to see a major earthquake, think that their building will not experience damage if one strikes, or prefer to address the problem after the fact. Not only building owners, but many Bucharest residents do not fear the next earthquake, with one-third believing there will "never be such an event in Bucharest."[19] In what is termed "loss bearing," people accept the burden and consequences of loss, either actual, potential, or perceived. It's a kind of coping mechanism for survivors and those at future risk. As Dragoş Toma-Danila and colleagues found, "In Romania, there is an entire generation that hasn't felt a major earthquake, does not know how to behave during one, does not conceive how considerable the damage could be or does not understand that immediate actions to mitigate the risk need to be taken (although many declare they are afraid of earthquakes)."[20] Some believe that if a building survived both the 1940 and 1977 earthquakes and the recent smaller ones, it must be resistant to earthquakes. This logic, of course, is flawed. A 2004 study with Bucharesters who had lived through at least one major earthquake found that only 10 percent feared the next one, and slightly more than half were "completely indifferent" to the risk.[21] Higher education was correlated with higher indifference. A majority of those in Bucharest will "take their chances" when faced with probable and possible future earthquakes.

As of January 2024, it is finally illegal to rent out or live in a building assessed into seismic risk Class I.[22] For decades the majority of people who lived in seismically unsafe buildings were those who did not see the risk or could not move, primarily the young and the old. For young people, buildings assessed into Class I, especially those that display a "red dot," were a low-rent option close to the city center, main university campus, and nightlife. For the elderly, many did not have the means, energy, or desire to move or repair. Our elderly neighbor was unhappy about the building's downgrade to Class I, but not because she feared its collapse. She believed that her apartment's property value would plummet if repairs were not approved or if authorities affixed a "red dot" to the building, and she lamented that her family's inheritance would be more of a burden than a benefit. She felt strongly that the apartment owners would never reach an agreement for the recommended repairs because the majority of them did not have the means to move during retrofitting, did not believe there would not be a strong earthquake in their lifetime, or did not care if there was.

When an Earthquake Strikes Romania: Warning and Preparation

The Romanian government, scientific community, and nongovernmental organizations have been working together to prepare for the next earthquake. The Romania Seismic Network (RSN) is Europe's first and the world's third earthquake early warning system. It uses detection equipment in more than 160 locations nationwide, a significant increase from 1977, when there was only one such device.[23] In the case of a significant disaster, the Interior Ministry's Emergency Situations Department (Ministerul de Interne, Departamentul pentru Situații de Urgență, DSU) is responsible for and oversees other entities engaged with disaster mitigation and management. In 2016, legislation created the National Committee for Special Emergency Situations (Comitetul Național pentru Situații de Urgență), housed in the Interior Ministry. This committee, alongside the prime minister, holds a coordinator role and leads the National Platform for Disaster Risk Reduction (Platforma Națională pentru Reducerea Riscurilor la Dezastre), which comprises scientific and technical experts as well as representatives of nongovernmental organizations, research institutions, professional associations, and the mass media. The General Inspectorate for Emergency Situations (General pentru Situații de Urgență, IGSU) is the primary organization for rescue operations and damage mitigation. It developed a post-earthquake response plan most recently tested in 2019 and its latest edition published in 2021.[24] The inspectorate, too, coordinates the National Disaster Risk Assessment at the National Level (Proiectul RO-RISK), a project to inform and coordinate with public administrators. Lastly, relevant state university academic departments (engineering, geophysics, construction, etc.) are tapped when needed.

At the time of an earthquake, Romanian scientists gather and analyze a wealth of data they share for disaster response. Since 2007 Romania has used ShakeMap, which creates real-time maps of ground movement and intensity from an earthquake. Since 2011, Romanian scientists have combined that data with structural damage estimates known as "SeisDaRo" and crowdsourced information from the "Did You Feel It?" campaign to quickly estimate possible building damage and potential casualties. This information is shared with first responders to triage their deployment. But it's not enough to depend on government action. Individuals can prepare, too (Appendix C).

While my family relocated to an apartment in a retrofitted and seismically safe building, that summer we were still living in Europe's most seismically at-risk capital city. There is no doubt that the risk to lives and property from Romania's next major earthquake will be significant, particularly in Bucharest. The Interior Ministry's most recent response plan estimates the potential for structural damage to residential buildings and resulting deaths in a range of hypothetical situations. Its worst-case scenario estimates anywhere from 838 to 33,569 residential buildings damaged in Bucharest, and potentially serious injury or death to as many as 45,000 people across the country. Others estimate that repair costs in Bucharest alone could be as high as $20 billion, with more than four hundred roads, covering almost twenty-five kilometers, damaged or blocked, and surface damage the equivalent of fifteen times the area of Bucharest's central park.[25] Today, Bucharest and Romania are better prepared than they were in the last decade of the communist era following the Ceauşescu regime's 1977 earthquake disaster response, and than they were twenty years ago, when a 2004 study concluded that "the population living in Bucharest is not prepared to cope with the consequences of a major earthquake."[26] Still, the potential damage, injuries, and deaths from the next strong earthquake will be substantial unless there are concerted efforts toward further mitigation, retrofitting repairs, and increased public awareness and consideration of the risk.

Notes

1 Administraţia Municipală, "Listă imobile expertizate."
2 For one collection of 1977 earthquake images, see White Mountain Property, "Bucharest 1977."
3 Direcţia Regională de Statistică, "Populaţia după domiciliu."
4 Hubbard and Timur, "He Warned Turkey about Earthquakes."
5 Toma-Danila and Armaş, "Insights into the Possible Seismic Damage," 1181.
6 E.-S. Georgescu, "Earthquake Engineering Development."
7 Ordonanţă privind măsuri pentru reducerea riscului seismic al construcţiilor existente, ordinance no. 20, January 27, 1994.

8 Order no. 770-6.173/NN, September 26, 1997.
9 Administraţia Municipală, "Listă imobile expertizate."
10 Administraţia Municipală.
11 Lege privind unele măsuri pentru reducerea riscului seismic al clădirilor, no. 212, July 12, 2022.
12 Ordin privind unele măsuri pentru avertizarea populaţiei din zonele expuse riscului la cutremur, no. 171, March 8, 2007.
13 Nicolae Noica, via Corina Leonte, email messages to author, October, 3, 4, and 11, 2023.
14 Regulament de organizare şi funcţionare a Comisiei centrale pentru prevenirea şi apărarea împotriva efectelor seismice şi alunecarilor de teren, June 6, 1996.
15 For a map, see Pre-Quake Project, "Riscul seismic al Bucureştiului."
16 Transparency International, "Romania Country Data."
17 Administraţia Municipală, "Programul Naţional de Consolidare."
18 Lege privind unele măsuri pentru reducerea riscului seismic al clădirilor, no. 212, July 12, 2022, article 10.
19 Armaş, "Earthquake Risk Perception," 1231.
20 Toma-Danila et al., "Earthquakes in Romania," 3204.
21 Armaş, "Earthquake Risk Perception," 1231.
22 Lege privind unele măsuri pentru reducerea riscului seismic al clădirilor, no. 212, July 12, 2022, article 5, amendments in effect January 3, 2024.
23 Toma-Danila and Armaş, "Insights into the Possible Seismic Damage," 1082; US Geological Survey, "Earthquake Early Warning."
24 Ministerul Afacerilor Interne, "Conceptia naţională."
25 Ministerul Afacerilor Interne, "Conceptia naţională," 7–8, 67; Bala and Toma-Danila, "The Strong Romanian Earthquakes," 31; Re:Rise, "Bucureştul după cutremur."
26 Armaş, "Earthquake Risk Perception," 1233.

Epilogue

Nicolae and Elena Ceauşescu took center stage in the regime's disaster recovery achievements in the wake of the 1977 earthquake. This was apparent inside and outside of Romania. The US Department of State noted that "the extent of the disaster drew all elements of the party and government leadership into relief and reconstruction activities, but it was Ceauşescu who directed and completely dominated the process. No other official was delegated meaningful authority or permitted to receive public recognition for his contribution."[1] At the earthquake's tenth anniversary, Radio Free Europe (RFE) broadcast:

> Everything turned into a huge show, "Ceauşescu and the Earthquake." On television and in the press, there was only him to be seen in scientifically studied and dramatized images. Among the ruins, Nicolae Ceauşescu was in his element. As the retrofitting and clearance work drew to a close, he began to enter through the basements, always preceded by photographers and the television cameraman, to give directions. He had become a specialist not only in miraculous rescues but also in cutting concrete, in the strength of materials and architecture. Those miraculously saved from under the ruins, many days after the earthquake, the wounded in hospitals, and those who had received a new home were thanking Him and Her.[2]

While history has moved beyond privileging the rulers and those at the center, in the case of the 1977 earthquake disaster response, given Ceauşescu's hold on power, we must start there. That does not mean that those with limited or even no access to that central power did not have agency to make significant contributions to the disaster recovery. They did. These figures included first responders: police and Securitate troops who dug out survivors and victims, firefighters who worked nonstop for days, and health-care workers who treated the injured and consoled victims' loved ones. They included scientific experts: specialists on the assessment commissions who immediately began the work of assessing the damaged buildings, then made plans and recommended evacuations to ensure safety; the workers who tried to complete as many structural repairs as

DOI: 10.4324/9781032632933-9

possible when issued an impossible deadline; and the engineer Gheorghe Ursu, who lost his life for sharing information about the public risk from the July cessation order that ended all structural assessments and hastened repairs. They included everyday Romanians: homeowners who made their repairs, the members of the diaspora who shared information and sent care packages, workers who clocked in for unpaid shifts and were pressured to donate portions of their salary and annual bonuses, women and children who worked in cleaning brigades, people who carried the wounded in their arms to hospitals and brought tea to first responders, and Bucharesters who called RFE with and for any information. As Rebecca Solnit explains, after a disaster people help one another address immediate needs, but because we are social beings, we express empathy, and a solidarity emerges that may not have been felt before the disaster.[3]

Only one week after the 1977 earthquake, RFE's director Noël Bernard lauded the solidarity of Romanians inside and outside the country as they connected to share information and help. He lamented, however, that it had taken a disaster to bring people together. He asked:

> What will the leadership do now, will they come to the level of the common man to repair and reconstruct the lives affected, or will they move along as usual, will they prioritize whether production levels rise or people's happiness?[4]

The 1977 earthquake disaster recovery activities the Ceauşescu regime prioritized demonstrate what it valued at the time: restoration of the economy and elimination of the "chaos" and "panic" that Ceauşescu saw. It clung to these values while disregarding public safety when it ended its disaster recovery well before the project's completion. This is the tragic legacy of the Ceauşescu regime's 1977 earthquake disaster response. Even with postcommunist risk mitigation efforts, a significant public risk in Bucharest remains. Many residents and visitors are unaware of it, or, if they are aware, they think it is overblown. "In Romania, there is an entire generation that hasn't felt a major earthquake, does not know how to behave during one, does not conceive how considerable the damage could be, or does not understand that immediate actions to mitigate the risk need to be taken."[5] This work aims to make one small contribution to change that.

Notes

1 Ashley and Munteanu, "Ceauşescu's Leadership Position, 544.
2 Tofan, *Antologia Radio Europa Liberă*, 280.
3 Solnit, *A Paradise Built in Hell*, 28.
4 Bernard, *Aici e Europa Liberă*, 100–102.
5 Toma-Danila et al., "Earthquakes in Romania," 3204.

Appendix A
The Civic Center Project and Its House of the People

Following the earthquake, until the Ceauşescu regime's end in 1989, the regime uprooted tens of thousands of residents, demolishing at least ten thousand homes as well as dozens of churches, synagogues, and public buildings, when it razed almost a quarter of Bucharest to build a new administrative and political complex, the Civic Center Project. The centerpiece of this project is the House of the People (Casa Poporului, see Figure A.1). Also once known as the House of the Republic (Casa Republicii), this building is today known as the Palace of the Parliament (Palatul Parlamentului). Still standing, it is the second-largest administrative and fourth-largest building in the world. The structure is 84 meters in height and has an area of 365,000 square meters; barely a third of it is used to house Romania's Parliament, a conference center, and three museums.[1]

Within weeks of the 1977 earthquake, Nicolae Ceauşescu resurrected the idea for such a complex from an exponentially smaller interwar draft plan, amending aspects of it to align with the city's 1972 systematization plan.[2] The 1977 earthquake and the regime's disaster response inspired Ceauşescu to start the project. There was little damage in the area cleared for its site. No evidence suggests that Ceauşescu was considering the Civic Center Project before the 1977 earthquake as part of Bucharest's modernization efforts. Yet, the earthquake, its destruction, and the recovery efforts led him to finally introduce it. The Civic Center Project's direct link to the 1977 earthquake is temporal: during the second week following the earthquake, Ceauşescu publicly announced the project's plan.[3]

Ceauşescu saw the Civic Center Project as the cornerstone of a larger urban renewal project. On March 22, 1977, he announced the plan at a private meeting with Political Executive Committee (PEC) members, architects, construction experts, and urban planners.[4] The plan would move all government and political offices, including the state archives, telecommunications center, and Romanian Communist Party (RCP) offices. Ceauşescu named its location in the Uranus neighborhood, near the city center, just as in the interwar plan.[5] He selected the neighborhood's highest point, Arsenal Hill, for the project's flagship: a grand building initially planned to host new offices for himself, the Central Committee of the Romanian Communist

Figure A.1 The fourth-largest building in the world, Ceauşescu's enormous House of the People, today known as the Palace of the Parliament, is located in Bucharest's Uranus neighborhood, demolished in the 1980s. Weeks after the 1977 earthquake, Ceauşescu announced his plan for the new building. Construction began in 1984 but was not completed during his regime. Photo by author, August 17, 2014.

Party (CCRCP), government ministers, the State Council, the Communist Youth Union, a large meeting hall, the state library, and a museum.[6] At the plan's announcement, Ceauşescu shared his vision for the area around the grand building: "There must be a large square ... and in the front of it, a terrace, because we must have a place to stand, we will make a semi-rotund meeting place."[7] He wanted a hall large enough to fit four to six thousand people, and a stadium, too, but not for soccer, because, as he explained, "we have enough of those, but for cultural and artistic festivals and large meetings." He recommended a hippodrome for parades, which was never built. The architectural style would not be modern (and specifically not cubist, he said), but would be "Romanian." He explained: "everywhere ... the architecture will have Romanian elements."[8]

The day after the private announcement of the Civic Center Project, state-run newspapers published a who's who of the meeting's attendees: "the most representative forces in the domains of architecture and building—researchers, university professors, builders from certain planning institutes and other construction organizations from the capital and other centers in the country."[9] It echoed Ceauşescu's claim of the earthquake's damage as justification, reporting,

> because certain important edifices were destroyed or severely damaged by the earthquake, or some very old buildings have shown to have low resistance and cannot be rebuilt, Comrade Ceauşescu presented to architects and construction engineers the issue of moving quickly to reconstruct certain important zones in Bucharest, and at the same time accelerating the activity to develop the capital of our patriotic nation.[10]

The project reflected the regime's ideological, Romanian nationalist, socialist agenda. It was to be a huge building, decorated with motifs of the many

Romanian regional communities, but not the other ethnic minorities—Hungarian, German (Saxon, Swabian), Roma, and so forth. Inspired still by his visit to North Korea and China earlier in the decade, Ceauşescu saw it as a monument, too, to himself. It served his vision of a capital city representative of his regime and its ideology. At its announcement, regime members signed on to the project and Ceauşescu's vision for it.[11]

Within two months of its announcement, architects presented Ceauşescu with three-dimensional models of the new Civic Center Project; a final plan followed that October.[12] In December, when Ceauşescu approved it, he declared: "I believe something beautiful and useful will be the result."[13] Anca Petrescu, a recent architectural school graduate, was the project's head architect. The design was not entirely hers, and the plans were never exactly final. In the 1980s Ceauşescu visited the construction site practically every Saturday and demanded changes in design or materials nearly every time.[14] Ceauşescu initially aimed for it to be built within five years, but its construction began in 1984, the year after he originally planned its completion.[15]

Building demolitions in Bucharest and across Romania following the 1977 earthquake through the regime's final decade were of several different types. First were those very few approved because of structural damage received from the earthquake. Second were those buildings Ceauşescu ordered destroyed just because, and not for safety or urban planning reasons. For example, in April 1977 the regime demolished the early eighteenth-century Ene Church, which sat in the heart of Bucharest.[16] It was one of the first nonresidential buildings the regime razed after the earthquake not because of structural damage, but because Ceauşescu demanded it. A month before the bulldozers razed the church, Ceauşescu described how the entire block where it sat would be demolished:

> We will take all from Ene Church Street to the hotel; all will be demolished and we will rebuild all of it, from the Dunărea Restaurant to the communal bathhouse … these will be in the first phase of the demolition plan.[17]

Third was the demolition of homes in rural areas with the continuation of the systematization policy. Last were those buildings cleared to make way for the Civic Center Project.

The Civic Center Project was not folded into the regime's urbanization policy; rather, it was considered a "special case" of urban planning.[18] In March 1977, Ceauşescu said that work under Bucharest's systematization plan was to be completed before any of the Civic Center Project construction began.[19] Mayor Ion Dincă, as head of the Bucharest city Executive Committee, oversaw the evictions and building relocations or demolitions for the neighborhood's destruction. Even before the project's start, people, of course, were well aware of the disruptions that would come from it. In August 1977 the Securitate reported that

> the idea of demolitions and building a new civic center area has created a state of agitation and anxiety, which is why some citizens have an inappropriate attitude, some even uttering insults to some departments of the state and the party.[20]

Starting in 1981 Ceauşescu signed several presidential decrees to appropriate the land, demolish existing structures, and uproot tens of thousands of people.[21] The neighborhoods destroyed—Uranus, Antim, and Rahova—were some of the oldest in the city, and they comprised mostly single-family homes and Bucharest's small Jewish quarter. The regime gave residents six months' notice of their home's demolition date, resettled them in apartments on the capital's outskirts, and compensated them at a fixed amount, two years of the average worker's salary, roughly 80,000 lei, roughly $4,000.[22]

In the mid-1980s, the regime stepped up the pace of neighborhood demolitions for the Civic Center Project but made efforts to preserve some buildings. Engineering experts picked up and rolled entire apartment buildings and churches.[23] Engineers slid Bucharest's Mihai-Vodă Church 300 meters using makeshift tracks, relocating it behind a large apartment complex, out of direct sight (presumably so Ceauşescu would not see it as he traveled through the capital).[24] The regime moved a few buildings and let one church and its bell tower remain, but it largely demolished structures in the Civic Center Project's way, including single-family homes, one- and two-story houses, small apartment buildings, and public buildings.[25] By the mid-1980s the Uranus neighborhood's destruction was well underway and had already displaced an estimated forty thousand people.[26] Some Romanians inside and outside the country, alongside international nongovernmental organizations, protested the demolitions, but with little impact. On June 25, 1984, more than seven years after the 1977 earthquake and the Civic Center Project's introduction, Nicolae and Elena Ceauşescu officiated over the House of the People's ground breaking.[27] That same year demolitions continued, including those of five Orthodox churches, among them the Vacareşti Monastery. In 1985 and 1986, the government demolished nine more churches, one synagogue, and the Romanian historian Nicolae Iorga's home.[28] During the 1980s, the regime mobilized as many as twenty-five thousand workers, many from the army, to move or demolish structures. Throughout its construction, the Civic Center Project was constantly behind schedule. It was delayed so much that it finished eight years after the revolutionary end of the Ceauşescu regime. There was little resistance within the regime to planning, demolishing Bucharest, and building the Civic Center Project.

Notes

1 For video footage, see Gage, "Memorialul Bucureştilor; Drcimage, "Veneţia din Suflete."
2 Sampson, *National Integration*, 82.

3 *Informaţia Bucureştiului*, March 23, 1977, 1.
4 Secţia Economică, 41/1977, 1–13, FCCPCR, ANR.
5 Deletant, *Ceauşescu and the Securitate*, 308; D.11.487, v. 2, 4, DB, CNSAS.
6 *Povestea Vorbei*, no. 195, 6-7, RFE/RL Romanian Broadcasting Department records, box 3811, folder 2, HILA; Secţia Economică, 41/1977, 3, FCCPCR, ANR.
7 Secţia Economică, 41/1977, 3, FCCPCR, ANR.
8 Secţia Economică, 41/1977, 3 rev. and 5–5 rev.
9 *Flacăra*, March 24, 1977, 3; *Informaţia Bucureştiului*, March 23, 1977, 1.
10 *Informaţia Bucureştiului*, March 23, 1977, 1.
11 Alexandru Budişteanu, pers. comm., August 2014.
12 *Povestea Vorbei*, no. 195, 7.
13 Secţia Propagandă si Agitaţie, 29/1977, 5, FCCPCR, ANR.
14 Secţia Cancelarie, 123/1988, FCCPCR, ANR.
15 *Informaţia Bucureştiului*, March 23, 1977, 1.
16 For images, see Galmeanu, "Filmul demolarii bisericii Enei."
17 Secţia Economică, 41/1977, 5 rev., FCCPCR, ANR.
18 Deletant, *Ceauşescu and the Securitate*, 304.
19 Secţia Economică, 41/1977, 4 rev. and 5, FCCPCR, ANR.
20 D.13.339, v. 37, 373, DB, CNSAS.
21 Secţia Cancelarie, 32/1981, FCCPCR, ANR.
22 Deletant, *Ceauşescu and the Securitate*, 295.
23 Danta, "Ceauşescu's Bucharest," 181.
24 *Adevarul*, "Inginerul care a mutat biserici."
25 Deletant, *Ceauşescu and the Securitate*, 311.
26 "Situation Report: Romania," November 2, 1984, HU OSA 300-8-47:45/4-16, p. 10, RFE/RL records, OSA.
27 For Romanian Communist Party footage, see Partidul Comunist Român, "25 iunie 1984 Inaugurarea"; Istorie Religie Civilizaţii, "Casa Poporului."
28 Deletant, *Ceauşescu and the Securitate*, 311.

Appendix B

The Identified 1977 Earthquake Victims

Compiled from archival sources, this list of the 1,308 identified victims of the March 4, 1977, earthquake, of the 1,578 presumed dead, is organized by where they died, listed by family and then given name. Shared names include an address to differentiate them, and foreigners have their home country in parentheses. Names (without diacritics) and locations are listed as they appear in the sources.[1]

Alba
Beleiu, Traian

Argeş
Alexe, Ion
Cercel, Stefan
Costache, Florea
Craineanu, Maria
Popa, Viorel Vasile
Voica, Dumitru

Bacău
Feldman, Sofia
Matei, Ioan
Rapas, Viorel
Sava, Constantin
Teslaru, Maria

Botoşani
Apetroaie, Adrian

Brăila
Aurica, Sanda
Manole, Vasile

Braşov
Bartos, Dominic
Cismas, Nicolae
Muresan, Ana
Silimon, Radul Bert Liviu

Bucureşti
Adam, Paraschiva
Albin, Hana
Alexandru, Stefan
Alterescu, Fany
Alterescu, Mendel
Ambrus, Elena
Ambrus, Letitia
Anastasiu, Filifteia
Andreescu, Aurelia
Andrei, Doina
Andrei, Elena
Andrei, Elena Cristina
Andrei, Nicolae
Angelescu, Maria
Angelescu, Mihai Dan
Angelescu, Mircea
Angelescu, Niculina
Anghel, Ana
Anghel, Felix
Anghel, Petre-Mircea
Anghelache, Grigore
Anghelescu, Gheorghe D. S.
Anghelescu, Mihail
Antoniu, Maria
Antoniu, Mihaela L.
Antoniu, Victor Paul
Apostol, Gheorghe
Apostol, Stefana
Apostolescu, Ana
Apostolescu, Armand
Arapu, Eugenia F.
Arghir, Ioana
Aron-Schwartz, Ana
Arseni, Paula
Arseni, Varvara
Atanasiu, Henry
Avasiloaei, Jenita

Avasiloaie, Gheorghe
Baba, Ionita
Baban, Gheorghe
Bacanu, Bogdan
Bacanu, Cristina Anca
Baconski, Anatol
Badea, Margareta
Badulescu, Ipsilante Ilie
Bagayamukwe, Walwimbo (Zaire)
Bakonsky, Clara
Balaban, Marioara
Balaban, Nicolae
Balan, Ilarie
Balanescu, Dumitru
Bals, Andrei
Baltaretu, Ernest
Baltaretu, Isabela
Baltaretu, Mariana
Baltateanu, Alexandra
Ban, Petru
Banaseanu, Julieta
Banga, Dumitru
Banga, Zamfira
Banica, Elena
Barasi, Antoneta
Bardan, Dumitru
Barna, Alexandru Lucian
Barna, Gloria
Barna, Vasile-Lucian
Baroi, Maria
Barth, Catharina
Basarabianu, Patru
Basturescu, Traian
Bazacliu, Maria
Bazgan, Paraschiva
Belicov, Anastasia Luiza
Belicov, Petre
Bena, Anca Rodica
Benescu, Marina
Bidilica, Maria Carmen
Bidiu, Luminita-Forina Mihaila
Bijutescu, Costel
Bilivschi, Vasile
Biolan, Aurica
Birladeanu, Maria

Birsan, Marta Elena
Birsan, Mircea
Birsanescu, Ileana
Birsanescu, Radu-Octavian
Bita, Ecaterina
Bobes, Maria
Bobes, Paul Teodor
Boboia, Rozalia
Boc, Doina
Bocanetu, Alexandru Catalin
Bociu, Aurel
Bociu, Cecilia
Bolat, Ioan
Bonciu, Balasea
Bonini, Dariu-Cristina
Bota, Teodor
Botez, Constantina
Bourceanu, Argripina
Bourceanu, Vasile
Bragari, Ligia
Bragari, Nicolae
Bragari, Parfenie
Bratu, Bianca
Bratu, Savin
Brintus, Ionel
Brintus, Rusanda
Brotman, Tirlea
Bruma, Ilarion
Bruma, Lidia
Bucsan, Elisabeta
Bucur, Aurel Grigore
Bucur, Laura
Bucur, Maria (Colonadelor 3)
Bucur, Maria (Poenaru Bordea 20)
Bucurescu, Alexandru
Bucurescu, Claudia
Bucuroiu, Georgeta
Bugulet, Sofia
Buh, Aglaia
Bulac, Florea
Bulai, Maria
Bulandra, Gheorghe
Bungete, Niculina
Burcea, Sevastita
Burchi, Maria

Burghiu, Ene
Burghiu, Lucia Sanda
Busch, Felicia
Busuioc, Ana
Busuioceanu, Ioan
Butur, Teodora
Buzincu, Jana Margareta
Buzoianu, Stanca
Caciula, Sorin
Calavreza, Emanoil
Calota, Stefania
Campeanu, Simion
Capata, Mihail
Capata, Stefania
Capatina, Adrian
Capatina, Alexandru
Capatina, Domnica
Capatina, Iosefina Rhea S.
Capatina, Silvana Michaela
Capatina, Theodor
Cappon, Emanuel
Cappon, Georgeta
Carafizi, Cristian
Caragiu, Toma
Carol, Corneliu Pericle
Carp, Lisetta
Carp, Saul
Carstea, Simona
Catut, Lucica Ana
Caurea, Daniela Ecaterina
Cazacu, Antonina
Cazacu, Eustatiu
Ceauşescu, Elena
Celea, Iuliana
Cerchia, Nicolae
Cerchia, Sofia
Cernea, Natalia
Cernomorcenco, Ana
Chear, Florea Florentina
Chele, Eugeniu
Chern, Avram
Chern, Hetty
Chicioroaga, Alda
Chicioroaga, Corneliu
Chiculita, Vasile

Chirca, Cezar
Chisiu, Nicolae
Chitic, Lenuta
Chitic, Petre
Chivu, Nicolae
Cimpeanu, Elena
Cimpeanu, Vladut
Cimporescu, Bogdan M.
Cimporescu, Cringuta Ileana
Cimporescu, Dionel
Cioaba, Constantina
Cioana, Marioara
Ciocan, Gheorghe
Ciociltan, Corina-Alexandra
Cioclea, Mihail
Cioescu, Florin-Neculai
Cirlan, Doina
Cirlan, Maria
Cirlan, Meita
Cirstea, Mircea Eugen
Cirstei, Agripina
Cirstei, George Roberto
Cirstei, Nicolae
Ciurea, Eufrosina
Coanda, Ana
Coban, Beatrice Victoria
Coban, Mihail
Coca, Petre
Cojocaru, Vasile
Coman, Theodor
Comsa, Elvira
Comsa, Vasilica (Emanuela)
Conea, Ana Dumitra
Constantinescu, Anoaneta
Constantinescu, Bogdan-G.
Constantinescu, Cecilia
Constantinescu, Victoria
Constantinescu, Viorica
Constantinidis, Panaiotis (Greece)
Constantinidis, Paraschevi (Greece)
Cordoneanu, Mihaela B.
Cordoneanu, Mihai Razvan
Cordoneanu, Stefanie Angela
Cornea, Alexandru
Cornea, Elena

Cornu, Pache Mihail
Cornu Maratea, Ioana-Maria
Corpodean, Ana
Cosoianu, Ioana Maria
Cosoianu, Ruxandra Sofia
Costa, Foru Olga
Costin, Elena
Cotau, Lapadat
Cotco, Silvia
Cotovitchi, Valentin
Covaci, Maria
Craciun, Grigorie
Craciunescu, Niculina
Craciunescu, Vasile
Cretoiu, Georgeta Aneta M.
Cretoiu, Stefan
Cretu, Atena Anca Nicoleta
Cringusi, Susana
Crisanu, Ecaterina
Cristea, Eugen
Cristea, Eugenia
Cristea, Vasile
Cristescu, Costel
Cristescu, Ecaterina
Cristescu, Niculae
Cristi, Boris
Cristodorescu, Elisa Beta M.
Crutescu, Beatrice Michette
Cucu, Olga
Cucu, Trandafira
Cucu, Ygnes Georgiana
Daboveanu, Elena
Damaschin, Dorin
Damoc, Adriana-Eliza
Dan, Deleanu Verona
Dan, Ilinca
Dankanits, Adam
Dankanits, Laszlo
Dariu, Aurora-Neti-Maria
Dariu, Gheorghe
Dascalu, Andrei
Dascalu, Elena
Dascalu, Gabriel
Dascalu, Paraschiva
Davidescu, Elena Olga V.

Decu, Ana
Dediu, Elena
Dediu, Stefan
Delian, Linorie
Denghel, Ioana
Diaconescu, Dana-Cristina
Diaconescu, Floarea
Diaconescu, Ioan
Diaconescu, Niculina G.
Dicu, Maria Hermina
Dimitrescu, George Anton
Dimitrescu, Gheorghe
Dimitrescu, Raluca
Dimitriu, Elena
Dinca, Angela
Dinca, Rada
Dinopol, Aurelia
Dinu, Dumitra
Dinu, Petre
Dita, Ion
Doaga, Margareta
Doaga, Radu
Dobre, Gheorghe
Dobre, Maria
Dobre, Stefan-Vasile
Dobrescu, Elena
Dobrescu, Ionela Florica S.
Dobrescu, Mihail Vasile
Dobrescu, Nicolae
Domnisor, Ecaterina
Dona, Dumitru Gabriel
Dona, Niculina-Gabriela
Donea, Gabriela
Dorneanu, Dragos
Dorneanu, Sorin
Dorobantu, Eleonora
Dosa, Marioara
Dragan, Sofia
Draghici, Lina
Draghioiu, Nicolae
Draghioiu, Rozalia
Dragoescu, Niculina
Dragomir, Maria
Dragomir, Marin
Dragu, Beatrice-Aneta

Dragu, Maria
Dragu, Viorica
Dragunoiu, Alexandru
Dragut, Ileana
Drath, Sofiea
Dumitrache, Ion
Dumitras, Alexandru Oct.
Dumitrescu, Constantin
Dumitrescu, Cornelia T.
Dumitrescu, Dumitru
Dumitrescu, Florian
Dumitrescu, George
Dumitrescu, Marioara
Dumitrescu, Paraschiva
Dumitrescu, Stefana
Dumitrescu, Tudor
Dumitrescu, Vasile
Dumitriu, Dumitra
Dumitriu, Maria
Dumitriu, Vasile
Dumitru, Gheorghe
Durlanescu, Aurel
Durlanescu, Verginia
Dutchievici, Nicolae
Dutchievici, Zoica
Eftimescu, Maria
Eftimie, Ion
Emanuel, Constanta
Emanuel, Eugeniu
Enache, Anca Mihaela
Enache, Gheorghe
Enache, Lucia
Enache, Olga
Enache, Petre
Enachescu, Marioara
Farcas, Floarea
Fer, Maria
Feteanu, Ion
Feteanu, Radu
Feteanu, Steliana
Fieraru, Ecaterina
Filipan, Maria Anastasie
Finchelstein, Anna
Finchelstein, Sergine Mihaila
Fintineanu, Dumitru

Florea, Caliopi
Florea, Maria
Florea, Marin
Florescu, Brindusa Anca
Florescu, Dinu
Florescu, Georgeta
Florescu, Gheorghe
Florescu, Margareta
Folescu, Ioan
Fraseniuc, Mihai
Fuia, George Cristian
Fuia, Hristu
Fuia, Vanghelia
Fussa, Ana
Gabrian, Avia
Gadarautanu, Delia
Gafita, Vlad Mihail
Gagyi, Maria
Galateanu, Cecilia Maria
Gatoschi, Coralia
Gatoschi, Elena
Gavrilescu, Emilia
Gazararian, Hasmic
Geamanu, Tudor
Georgescu, Alina Ioana
Georgescu, Cecilia
Georgescu, Constanta
Georgescu, Maria
Georgescu, Niculai
Ghenu, Eugenia
Ghenu, Mihai
Ghenu, Mihail
Gheorghe, Maria
Gheorghe, Oprea
Gheorghe, Serban
Gheorghe, Teodora
Gheorghiade, Adriana
Gheorghiade, Camelia
Gheorghiu, Dinu Serban
Gheorghiu, Ecaterina
Gheorghiu, Ecaterina Rodica
Gheorghiu, Florica
Gheorghiu, Gheorghe
Gheorghiu, Herta
Gheorghiu, Mircea

Gheroghiu, Maria
Ghersanoc, Iosif
Ghersanoc, Mina
Ghimici, Mina
Ghinea, Ion
Ghinea, Maria
Ghingold, Liba
Ghinoiu, Maria
Ghinoiu, Tudor
Gijila, Mihaela Cristiana
Gioga, Aurica
Glavu, Ilie
Glavu, Irina
Glicor, Valeriu
Goaga, Gheorghe
Goaga, Radina
Goldemberg, Tipra
Goldemberg, Vily
Golianu, Aurelia
Golianu, Dan
Grabovschi, Elena
Grabovschi, Laurentiu
Grajdeanu, Cleopatra
Granet, Iulian
Granet, Silvia
Grigorescu, Rucsanda
Grigoriu, Aglaia Maria
Grigoriu, Livia Georgeta
Grigoriu, Lucia
Grigoriu, Vasilica
Griscenco, Esfir
Grodinischi, Teodor
Gross, Helene
Gruber, Gabriela Irma
Has, Anastasia
Heberling, Florica Margareta
Heberling, Ion
Heisser, Somoil
Hening, Friederich
Hening, Melitta
Hening, Susana Hermaine
Hobeanu, Teodor Mihail
Hobeanu, Teodora Georgeta
Hondru, Antonina
Horceag, Ana

Hristea, Ioana
Hutanu, Elena
Hutanu, Mihai
Hutanu, Vasile
Iacob, Gheorghe
Iacobini, Maria
Iatan, Marioara
Iatan, Mircea
Ilie, Gheorghe
Ilie, Ioana
Ilie, Marcela Iuli
Iliescu, Vasilica
Iliev, Ioana
Ioan, Catalin-Marian
Ioan, Georgeta Ivonne
Ioanid, Maria
Ioanitescu, Andrian
Ioanitescu, Florica
Ioanovici, Nicolae
Ion, Florin
Ion, Violeta
Ionescu, Adrian
Ionescu, Beatrice
Ionescu, Constantin
Ionescu, Cristina
Ionescu, Eufrosina
Ionescu, Gabriela Constranta
Ionescu, Ioana
Ionescu, Letitia Maria
Ionescu, Maria
Ionescu, Matrona
Ionescu, Teodor
Iordachescu, Barbu-Costin
Iordachescu, Emil-Constantin
Iordachescu, Maria-M.
Iordachioiu, Rozalia
Iorgulescu, Adriana
Iorgulescu, Luminita Ileana
Iosif, Elena
Iovanescu, Dan
Isacescu, Mihai
Isserson, Lucia
Istocescu, Dumitru
Istrate, Gheorghe
Istrati, Elpinichi-Despina

Ittu, Teodor
Ittu, Virginia
Ivasiuc, Alexandru
Joldes, Sofia
Katz, Ada Simona
Katz, Allan Oded
Kirmayer, Cristian
Kirmayer, Ruth
Klang, Rasel
Laptoiu, Catalin
Laptoiu, Daniela
Laptoiu, Grabriela
Laptoiu, Ion
Lazarescu, Aneta
Lazariciu, Anca Ioana
Lazariciu, Eugenia Laura
Lazaroiu, Vasilichia
Lazarovici, Florin
Lazarut, Victoria
Ledca, Vergil
Lejnenco, Valentin
Lica, Ion
Ligda, Julieta
Ligda, Maria
Linguraru, Stefania
Liteanu, Adriana
Livescu, Marie Claire
Luca, Ovidiu Dan
Lugojanu, Constantin
Lugojanu, Maria
Lupu, Vasile Costel
Luta, Maria
Magda, Aurel
Maiescu, Nicolai
Maiescu, Silvia
Man, Dumitru
Manac, Emilia
Manac, Marin
Manailescu, Gheorghe
Mancas, Mari Jana
Mancas, Romana
Manciulescu, Petru Stefan
Manciulescu, Rodica
Manea, Evghenia
Mantz, Constanta

Mantz, Ovidiu Dan
Maracineanu, Ioana Mihaela
Maran, Anna
Marcu, Arllette
Marcu, Haim
Marcu, Karly
Marcu, Lucian-Valeriu
Marcu, Minerva
Marcu, Tilla
Mardiros, Ion
Mardiros, Zamfira
Marescu, Catoni
Marescu, Paulina
Margineanu, Alexandru
Margineanu, Aurelian
Marin, Alla
Marin, Marin
Marinescu, Constantin
Marinescu, Dan Virgil
Marinescu, Elisabeta
Marinescu, Irina
Marinescu, Margareta
Marinescu, Nicolae
Marinescu, Sanda
Marinescu, Vasile
Martin, Constantin Traian
Martin, Lucica
Martin, Paraschiva
Matache, Elena Constanta
Matache, Ion
Mateescu, Carmen Rodica
Mateiu, Marin
Medrea, Iuliana
Meitoiu, Elena
Merisescu, Virgil
Metta, Constantin-Serban
Micia, Aurelia
Micu, Ioana
Midoiu, Tudorita
Miereanu, Bogdan
Miereanu, Constanta
Miereanu, Ion
Mihaescu, Ruxandra
Mihai, Doina
Mihai, Gheorghe

Mihaila, Gheorghe
Mihailescu, Octavian
Mihailescu, Penelope
Mihailovici, Louis Robert I.
Mihalcea, Ana
Mihaly, Anna
Mihu, Alexandru Andrei
Mihu, Ileana
Mija, Dionisie
Mija, Maria
Mileschin, Alexandru
Milica, Dan
Milica, Dragos
Milica, Elisabeta Virginia
Milica, Stan
Miloglav, Maria
Minea, Constantin
Mircescu, Valentin
Mirea, Ion
Mirea, Nicolae
Mirescu, Valeria
Mirica, Bogdan
Mirica, Iuliana
Miroiu, Sara
Miroiu, Traian
Mitranescu, Gheorghe
Mocanu, Georgeta
Mocanu, Isabela Cristina
Modrogan, George
Moga, Ana Teodora
Moga, Ioan
Moisan, Dumitru
Moisescu, Alexandrina
Moisescu, Georgeta Stefania
Moisescu, Mircea
Moldovan, Ana
Moldovan, Cristina
Moldovan, Rafila
Moranescu, Silvia
Morocos, Nicolae
Moruzov, Stefana
Mosteoru, Maria
Motateanu, Dumitra
Motriuc, Mihai
Muhlrad, Ides

Munteanu, Aurelia
Munteanu, Gabriela Victoria
Munteanu, Gyorgy
Munteanu Lucinescu, R.-M.
Muntianu, Alexandrina
Muresanu, Ioan
Musetescu, Adriana
Musetescu, Bogdan Cornel
Mustaciosu, Maria-M.
Mustaciosu, Valeriu Dumitru
Mustata, Alexandrina E.
Mustata, Dumitru
Nadolu, Dana
Nafornita, Cristiana Ofelia
Nafornita, Dimitre
Nagy, Irina
Naiculescu, Constanta
Nanescu, Margareta
Nanescu, Sorin
Nanuveanu, Maria
Nedelcu, Petre
Nedelcu, Smaranda
Negoi, Eleonora
Negoi, Marina
Negomir, Ana
Negrini, Paraschiva
Nemes, Teodora
Nemesu, Gheorghe
Nenitescu, Elena Ana
Netta, Secareanu Lucia
Nica, Maria
Nica / Nicu, Anton
Nica / Nicu, Elisabeta
Nica / Nicu, Oana
Nica / Nicu, Onut
Nicolae, Emilia
Nicolae, Ofelia Carmen
Nicolaescu, Lucia
Nicolaescu, Nicolae
Nicolaescu, Silvia
Nicolaescu, Vasile
Nicolau, Constantin
Nicolau, Demostene
Nicolau, Ileana
Nicolau, Ion

Nicolau, Irina
Nicolau, Petre
Nicolau, Sofia
Niculescu, Ilie
Niculescu, Silviu
Nita, Margareta Mariana I. L.
Nita, Violeta
Nitu, Luciea
Nuhamsohn, Herman
Nuhamsohn, Maria
Nutu, C. Florica
Oance, Victoria
Olaneanu, Dumitru Stelian
Olescu, Florina
Olescu, Victoria
Oncescu-Bestelei, Ana
Oncescu-Bestelei, Lucian
Oncescu-Bestelei, Mihai
Onete, Vasile
Oprea, Elena
Opris, Alexandru
Oprisan, Constantina
Oprisescu, Simona
Oravanu, Jean Sirius
Ostermaier, Elisabeth
Paltineanu, Elisabeta
Panaitescu, Elvira
Panaitescu, Viorica
Pandele, Bogdan Ionut
Panturu, Dumitru
Panturu, Vasilica
Papa, Dimitriu Marta
Papa / Papadimitriu, Dimitriu Petru
Papadima, Georgeta Victoria
Papadis, Tasos
Papazisi, Mihaela Rita
Paraschiv, Alexandru
Paraschivescu, Mircea
Paraschivescu, Neonica
Pasca, Georgeta Aurora
Patrascu, Liliana
Patrascu, Liviu-Florin-G.
Paul, Vasilica
Pavelescu, Ana
Pavelescu, Vasile

Pavelescu Segarcea, Aurelia
Pavlovschi, Ana
Pelmus, Ecaterina
Pelmus, Victor Gheroghe
Peris, Onoriu Mircea
Petis, Victor
Petre, Cecilia
Petre, Mihaela
Petrescu, Alexandrina
Petrescu, Aurica
Petrescu, Gheorghe Romulus
Petrescu, Maria
Petrescu, Matilda
Petrescu, Paul
Petrescu, Victoria
Petrila, Virgil
Petrov, Jorgyia
Petrusca, Marioara
Piciorus, Eugenia
Pintea, Horia Constantin
Pintilie, Gheorghe
Pintilie, Ruxandra
Pirici, David
Piroska, Ioan
Piso, Ioan
Pituleanu, Elena
Platz, Iohan
Platz, Valeria
Pop, Felix Max
Pop, Suzana-Cecilia
Popa, Adriana
Popa, Ion
Popa, Liviu
Popa, Lucea
Popa, Mariea
Popescu, Alexandru Valentin
Popescu, Anca Marina
Popescu, Daniela
Popescu, Daniela-Rene (Poland)
Popescu, Dorina (Poland)
Popescu, Elena (Calea Griviţei 23)
Popescu, Elena (Intrarea Horbotei 8)
Popescu, Eugenia
Popescu, Eugen Pompiliu
Popescu, Fanel-Stelian

Popescu, Gabriel
Popescu, Georgeta
Popescu, Gheorghe
Popescu, Ion
Popescu, Lucia
Popescu, M. Corneliu
Popescu, Marcela
Popescu, Maria
Popescu, Maria Domnica
Popescu, Mircea
Popescu, Rodica
Popescu, Sebastian Ion
Popescu, Stanca
Popescu, Stefania
Popescu, Stela
Popescu, Teodor
Popescu, Vasile
Popescu, Vasilica
Popovici, Renee Emilie S.
Popp, Maria
Preda, Ion
Preda, Veronica
Racman, Dragos Alexandru
Radescu, Maria
Radescu, Mircea
Radescu, Silvia
Radoi, Aurica
Radu, Alexandru
Radu, Dumitru
Radu, Ecaterina
Radulescu, Alexandrina
Radulescu, Anne Aymone
Radulescu, Chirita
Radulescu, Floarea
Radulescu, Georgeta L.
Radulescu, Ilinca
Radulescu, Maria
Rainhart, Cristina Iulia
Rameta, Dan Rasvan
Raux, Maria-Mariana
Reiner, Alfred
Reiner, Mircea
Reiner, Stefana
Remus, Aurel
Rodean, Ion

Român, Aurel
Român, Maria
Rosu, Elena
Rosu, Gabriela
Rosu, Iulia
Rosu, Traian
Rotman, Dora
Rotman, Iosif
Rozanide, Constantin
Rozenfeld, Alexandrina
Rozenfeld, Irina
Rozenfeld, Mihaly
Rudaru, Gheorghe
Rudeanu, Ion
Rufa, Maria
Rus, Ion / Ioan
Ruse, Polixenia
Rusescu, Margareta V.
Russo, Ritta
Sachelarie, Anca Maria
Sandru, Nicolae
Sandu, Alexandrina
Sandu, Alexandru
Sandu, Emilia
Sandu, Marian
Saraga, Samoil
Sarba, Ecaterina
Sarkadi, Iulianna
Sasareanu, Paraschiva
Saulea, Mihnea Radu
Sava, Aristita
Sava, Gabriela
Sava, Ifrim
Savu, Verona
Scally, Romvald Teodor E.
Schifner, Leonard
Schiopu, Maria
Schiopu, Vasile
Schrager, Savel
Schuman, Ecaterina
Scintei, Alexandrina
Scintei, Cristina-Mihaela
Scotnitchi, George
Scurtu, Maria
Secareanu, Vasile

Sechel-Mandoi, Florica
Secui, Virgiliu
Serban, Petre-Bogdan
Serbanescu, Florica
Seropian, Armenuhi
Seropian, Daniela Silvia
Sezonov, Ana
Siadbei, Ion
Siadbei, Matilda
Siboriu, Alexandru
Sichitiu, Mihail
Simioana, Florian
Simion, Daniela
Simion, Ionel
Simion, Maria
Simu, Gheorghe
Sinca, Niculina
Sirbu, Zina
Sitaru, Edith Therese
Sitaru, Jean
Slavinschi, Alexandru
Slobozeanu, Gheorghe
Smeu, Andrei Claudiu C.
Smeu, Traian
Smeu, Traian Bogdan
Socolov, Nicolae
Socolov, Valentina
Sofroni, Vera
Soigan, Ileana
Soigan, Mircea-Florian M.
Solomon, Karoline
Somesan, Ion-Florin
Sonea, Ioan Emil
Sonea, Margareta
Sora, Domnica
Spataru, Stefan Mihai
Spataru, Susana
Stainberg, Izac
Stainberg, Roza
Stamatin Iovanescu, R.L.
Stan, Constantina
Stanca, Letitia Valeria
Stanca, Octavian
Stanca, Roxana M.
Stancu, Maria

Stanculescu, Anca Maria
Stanculescu, Emilie Irena
Stanescu, Ioan
State, Ion
Steanta, Elena
Stefanescu, Anca Ruxandra
Stefanescu, Camelia A.
Stefanescu, Elena
Stefanescu, Florica
Stefanescu, Lylliana
Stefanescu, Marcel
Stefanescu, Maria-Emilia
Stefanescu, Marioara
Stefanescu, Nicolae
Stefanescu, Razvan
Stefanescu, Stefan
Stefanescu, Varvara P. O.
Stefanescu, Werner H.
Stefanescu, Zanoaga E. A.
Stefanita, Elena
Stegaru, Victoria
Stinghie, Angela
Stirbu, Maria Gina
Stirbu, Paul Stefan
Stoenescu, Sultanica
Stoian, Ion
Stoica, Livia
Stoica, Nicolae
Stoica, Valentin
Stoica, Vasile
Stoicescu, Efrosina
Stoicescu, Razvan-Mihai
Straus, Dumitra Ecaterina
Strava, Mircea-Ioan
Stroe, Dumitru
Stroe, Lidia
Stroe, Stefania
Stuparu, Florica
Suleimanovici, Sadin
Susan, Nicolae
Szilagyi, Geza-Attila
Tabacaru, Eufrosina
Tailoru, Vasile
Tamler, Ernst
Tanase, Cristian

Tanase, Florica
Tanase, Gelu Octavian
Tanase, Gheorghe
Tanase, Maria-Magdalena
Tanasescu, Eugenia
Tanasescu, Lucia
Tanska, Hanna (Poland)
Tanska, Wieslawa (Poland)
Tantulea, Gherghina
Tapu, Constanta
Tapu, Marina
Taranu, Aurel-Emil
Taranu, Aurelia
Taranu, Cosmin
Taranu, Marian Petre
Taranu, Wanda Marina
Teaca, Dumitru
Teaca, Maria
Teodor / Teodoru, Aurelia
Teodor / Teodoru, Dionisie
Teodorescu, Mihai
Teodorescu, Rodica
Teodoru, Aurelia
Teodoru, Ecaterina
Teodoru, Elena
Teodoru, Victor
Tepelus, Petre Constantin
Teposu, Virginia
Terchila, Carmen Ana
Theodosiu, Niculae
Tigoiu, Mircea
Tigoiu, Simona Florenta F.
Tipa, Gheorghe
Tiru, Barbara
Tisu, A. Constantin
Tomescu, Vasile
Toncea, Angela
Torok / Turuk, Etelka
Trandafir, Silvia
Trica, Florea
Trocan, Constantin
Tuculescu, Andreia Voica
Tudor, Petre
Tudor, Stefan
Tudorache, Vasile

Tugui, Florica
Tugui, Ionel
Tuinea, Lina
Tutoiu, Didina
Udrica, Antoaneta
Ungureanu, Marius Stefan
Ursescu, Elena
Ursescu, Vasile
Ursescu, Virginia
Vaduva, Alexandru
Vaduva, Ilie
Vaduva, Niculina
Vaduva, Stelian
Vagner, Matilda
Vahnovann, Valeria
Valerian, Mihail
Vancea, Livia
Vas, Andriana Mira
Vas, Marta
Vas, Pavel
Vas, Petre Lucian
Vasile, Alisa
Vasile, Ion
Vasile, Odeta Ioana
Vasilescu-Dinu, Adrian-Ion
Vasilescu-Dinu, Antoaneta
Vasiliu, Dumitru
Vasiliu, Eugen
Vass, Marta
Vatamanu, Alexandrina
Vatamanu, Nicolaie
Veis, Erna
Venczel, Vanda
Veres, Zoltan
Vermont, Cornelia
Vermont, Teodor
Vezeteu, Alexandru
Vezeteu / Vizitiu, Ana M.
Vilceanu, Alexandrina
Vinatoru, Carmen Adriana
Vinatoru, Paul George
Visan, Maria
Visinescu, Rodica G.
Vizante, Viorica

Vlad, Nicolae
Vlase, Andrei
Vochin, Virgina
Voin, Nicolae
Vrinceanu, Carmen
Vrinceanu, Daniela
Vrinceanu, Elisabeta
Vrinceanu, Nicolae
Weber, Carol
Weber, Mica
Weis, Oloze Constanta
Wozniak, Andrzey R.
Wylezynski, Alexandrina
Wylezynski, Kazimierz
Zaharia, Elena
Zamfir, Ioana
Zamfiriu, Gheorghe
Zapata, Solbalvarro R. B.
Zednic, Gheze Victor
Zimel, Herman
Zorzon, Paula

Buzău
Baraboiu, Voica
Caraman, Horia
Caraus, Luca
Dumitru, Aurel
Georgescu, Elena
Grigore, Nicolae
Ispas, Maria
Lazar, Cornelia
Moldoveanu, Stefana
Stroe, Alexandru

Călăraşi
Bogdan, Ioana
Miscu, Petrica
Stefan, Carmen Iuliana

Cluj
Pastean, Ioan Romulus
Serdan, Alexandru
Verde, Nicolae

Constanţa
Caraman, Despina
Dinca, Fanita
Grabcenco, Camelia
Grabcenco, Mihai
Grabcenco, Ovidiu-Geo
Popescu, Elena
Popescu, Gheorghe
Stoica, Maria

Covasna
Facaianu, Maria

Craiova
Andrei, Victoria
Brincus, Daniel
Bunica, Tudor
Chiriacescu, Adelina-E.
Ciocan, Marin
Ciochie, Floarea
Cunita, Silvia
Dobre, Florica
Dragomirescu, Maria
Fieraru, Ioana
Gheroghe, Silvia
Gherorghiu, Radita
Ghita, Maria
Grecu, Eduard
Grecu, George Daniel
Grecu, Manuela
Iacob, Nastasia
Ionescu, Ion
Isaila, Radu-Viorel
Jega, Maria
Musiu, Radu
Nedelcu, Constantina
Pirvu, Maria
Predescu, Maria
Roman, Elena
Teaca, Margareta
Ungureanu, Angela
Vulpe, Stefan
Zorenghea, Pastita

Dâmboviţa
Anahorlis, Dumitru-Const.
Coman, Ioana
Ghita, Marin
Negulescu, Constantin
Osman, Paul
Popescu, Petre
Soare, Marin

Deva
Vintila, Ioan
Dobrogea
Chirani, Stelian

Dolj
Cristescu, Alexandru
Gherea, Filuta
Ginzoiu, Anica
Lilea, Ion
Mitroi, Petre
Negoe, Nuta
Rosu, Ionel
Senoaica, Florica
Simion, Stefana
Smeu, Ana Paulina
Staicu, Maria
Vaduva, Liviu Mihail
Vulpe, Lidia

Focşani
Plesa, Sava

Galaţi
Barbu, Mimi
Chiculita, Ilinca
Cirjanu, Leontina

Giurgiu
Dionisie, Codruta-Maria

Glina
Gheorghe, Maria

Gorj
Pirici, Vasile
Simu, Anisia
Simu, Ion

Hunedoara
Crocos, Gabriela Stancuta
Crocos, Mihaela Codruta

Ialomiţa
Cucu, Elena
Iorga, Alexandrina
Marin, Stan
Marina, Gheorghe
Oprea, Ion
Tudorache, Ionut

Iaşi
Bivol, Sergiu
Dobre, Ilie
Ieftinie, Vasile
Micusca, Costache

Ilfov
Barbu, Gheorghe
Bucur, Vasile
Butu, Lica
Calin, Maria
Calin, Marin
Danila, Marita
Dinica, Ioana
Dinica, Maria
Draghici, Mihail
Dragnea, Maria
Dumitrescu, Avram Ion
Dumitru, Ion
Duna, Vasilca
Francu, Ioana
Georgescu, Adina Gabriela
Grama, Steluta
Horescu, Dumitru
Ion, Mariana

Ionita, Floarea
Marin, Constantin
Marin, Gheorghe
Matache, Gheorghe
Mihaita, Floarea
Muresan, Vasile
Nica, Elena
Paraschiv, Maria
Paraschiv, Marin
Pateanu, Aurica
Patrascu, Florea
Penciu, Aurelian
Popa, Victoria
Popescu, Florica
Preda, Gheorghe
Radu, Maria
Sandu, Tudora
Stefan, Gheorghe
Stoica, Victoria
Tabla, Aurica
Tanase, Angela
Tanase, Radu
Tudorache, Florina
Udrea, Ion
Zamfir, Gheorghe

Mehedinţi
Pletea, Marius-Constantin

Mureş
Piso, Magdalena-Iohana

Neamţ
Chiribau, Olga
Ghirvu, Elena

Olt
Badea, Adriana
Bocai, Florea
Sirbu, Mircea
Piatra Neamţ
Luca, Sofia

Piteşti
Cojocariu, Vasile
Jamba, Anatolie (Congo)
Popescu, Radu-Leon

Ploieşti
Babau, Ion
Dumitru, Vasile
Enescu, Steliana
Gheorghe, Ana
Ilinca, Constanta
Ionescu, Filofteia
Maruntis, Adrian
Mihaila, Maria
Nitulescu, Aurica
Pantilie, Elena
Schneider, Ilinca
Stegaru, Irina
Ungureanu, Constantin
Ungureanu, Emilia

Prahova
Calugareanu, Vasile
Carpen, Aurelia
Ciuta, Constantin
Gonciu, Constantin
Ionescu, Cecilia
Lupu, Stefan
Marta, Petru
Minoriceanu, Nicolae
Neagoe, Maria
Neagu, Ion
Nia, Vasile
Nicoara, Elena
Nicoara, Ghita
Pana, Petre
Petre, Constantin
Pirtoaca, Maria
Tambalagiu, Nicolae
Tatu, Ecaterina
Teodorescu, Iulia
Vasile, Constantin
Vasile, Maria

Zubas, Ion
Zubas, Maria

Sălaj
Bene, Rodica Ana
Fer, Nicolae Jisus

Sibiu
Terchila, Valeriu

Suceava
Bintu, Niculai
Cazac, Constantin
Spartariu, Ion
Teodorescu, Ferdinand

Târgovişte
Volosciuc, Maria
Volosciuc, Mariana
Volosciuc, Vasile

Teleorman
Bacau, Marioara
Birlogeanu, Stefan
Blaga, Partenie
Ceauşescu, Afrodita
Craciunescu, Sergiu-Iulian
Dumitrescu, Alexandra
Grigorescu, Lucian Lucreţiu
Iordachioaia, Mariana
Olteanu, Victor
Pesu, Marian
Raducanu, Mihaela
Streche, Ioana
Streche, Mariana
Streche, Tudorel
Udrica, Niculina
Vlad, Cristian
Vlad, Gheorghe
Vlad, Ilie
Vlad, Lina

Timişoara
Adam, Ingeborg
Catrina, Gheorghe
Stefanescu, Romeo-Mircea

Tulcea
Ilie, Andrei
Vladimirov, Dumitra

Vaslui
Drutu, Catalina
Drutu, Ecaterina
Filip, Ioan
Florea, Ecaterina
Ionita, Corneliu
Micu, Danut
Zaharia, Emilia

Vîlcea
Locksmithu, Filofteia
Palmes, Dorin

Vrancea
Chirila, Ruxandra
Dulhan, Ion
Mocanu, Ion
Negrea, Dorinel
Vrinceanu, Maria

Zimnicea
Arapu, Constantin
Gheorghe, Ortansa
Ioardache, Stan
Pietreanu, Gheorghe
Pietreanu, Petre
Soare, Tudor

Note

1 Secţia Administrativ-Politică, 12/1977, 2–53 rev., FCCPCR, ANR; D.14.800, v. 1, 14–78 rev., DB, CNSAS.

Appendix C
An Earthquake Preparation Guide for Romania

In seismically active areas no one is ever completely safe from an earthquake. There are, however, specific, tangible actions one can take to prepare. It is important to learn about the factors that increase one's risk and ways to reduce it; connect to alert systems to be informed in the case of an earthquake or other disaster; and make a plan for yourself, your loved ones, and your things.

Step 1: Learn about Risk

Many factors contribute to damage, injuries, and death from an earthquake. Its measured strength, depth, wave, and vibrational characteristics are primary factors, but the time of day it occurs and the distance from its epicenter also contribute to the scope of damage to built structures and the resulting injuries and deaths. The largest risk is from damage caused by large, seismically unsafe structures: buildings, transportation systems, bridges, roads, and so forth. It's imperative to understand the potential risk in the neighborhood where you live, work, or plan to visit. In Bucharest, the city center and the neighborhoods in Sector 1 and Sector 5 present the highest risk, where more than three-quarters of all buildings built before 1940 do not meet current seismic building codes. Consider viewing the digital and video tour "Bucharest and Earthquakes" put together by the National Research and Development Institute for Geophysics (Institutul Naţional de Cercetare-Dezvoltare pentru Fizica Pământului, INCDFP).[1] You can also look up whether specific buildings where you live, work, or visit have been assessed for seismic stability and, if so, what seismic class or urgency level they are assigned. Bucharest's Municipal Administration for the Consolidation of Buildings with Seismic Risk (Administraţia Municipală pentru Consolidarea Clădirilor cu Risc Seismic, AMCCRS) hosts activity reports as well as a list of buildings assessed using the seismic risk classification system.[2] If you own or live in a building assessed as seismically unsafe or one that is not up to seismic code, see Chapter 7, which lists government agencies to apply to for assistance.

Step 2: Connect to Alert Systems

Earthquakes have extremely short or even nonexistent warning times. Unlike hurricanes, snowstorms, heavy rainfall, excessive heat, unhealthy air quality, biological or radioactive contamination, or other potentially deadly events where warning and preparation times can be measured in minutes, hours, or days, an earthquake's warning time is measured in seconds. Seismometers measure the earth's movement in waves, with an earthquake creating two. The P wave (primary or compressional wave) occurs first. The S wave is slower than the P wave; it arrives second and usually represents stronger earth movement. The time between the two is usually only a few seconds and the only opportunity for an alert warning of the stronger movement to come. These few extra seconds could mean the difference between life and death, or serious and minor injury. Romania has a sophisticated early warning system that tries to exploit this seconds-long potential warning window. The INCDFP hosts the Rapid Earthquake Early Warning System (REWS) and shares information on the internet (http://ews.infp.ro/), via social media, and the smartphone app EarthBot.[3] The Department for Emergency Situations (Departamentului pentru Situaţii de Urgenţă, DSU), the General Inspectorate for Emergency Situations (Inspectoratul General pentru Situaţii de Urgenţă, IGSU), and the local Territorial Inspectorates for Emergency Situations (Inspectoratele Teritoriale pentru Situaţii de Urgenţă, ITSU) disseminate information through the REWS alert system not only for earthquakes but for extreme weather, floods, terrorism, and other threatening situations. The scientists who helped develop it state simply, "REWS can save your life."[4] In Bucharest, depending on the distance from its epicenter and the depth of the earthquake, the notification can come twenty to twenty-five seconds before one feels the secondary, more dangerous, wave. Even if one viewed the alert exactly at its arrival, twenty seconds is most likely not enough time to evacuate a building. It might be enough time, however, to take cover, or, if outside, to move away from large structures.

Romanian Warning and Alert Systems
- Telegram: @alertacutremur_bot
- X: @cutremurinfo, @rews_infp, and @incdfp
- Internet sites: http://ews.infp.ro/ and https://infp.ro/

Step 3: Make a Plan and Get Organized

To help people across Romania understand what they can do to prepare for different potential disasters and how to respond if they occur, the DSU hosts the website *Be Prepared* (Fiipregatit.ro), a public-private collaboration between the government's Urgent Situations Inspectorate (Inspectoratul General pentru Situaţii de Urgenţă, IGSU) and the nongovernmental organization Asociaţia CivicNet. On the Fiipregatit.ro website, one can learn how to prepare for and respond to floods, severe storms, tornadoes, fire, earthquakes, pandemics (as it did for COVID-19), and even domestic violence. In the case of earthquakes, it links to general and real-time global and domestic seismic information from the INCDFP.[5]

Regardless of your building's structural stability, it is important to secure large furniture pieces (wardrobe, dressers, bookcases, cabinets, etc.) so they are less likely to fall during an earthquake. Since the warning times are in seconds and evacuation is highly unlikely, prepare and practice what you would do and where you would go to protect yourself from falling items or debris. Inside homes, the kitchen and bathroom may be the safest areas. It's important to consider how to find cover under furniture, like a table or bed, or inside furniture, like a dresser or cabinet. The "Be Prepared" public information campaign recommends that people complete several preparation activities: pack one backpack per family member with water and food for three days, first aid supplies, cash, phone charger, and medications; gather work, school, and residential addresses and phone numbers, family names and contact information; and arrange and share one or more meeting places to reconnect if separated.

In addition to the *Be Prepared* site, the DSU launched the public interest campaign "Don't Shake during the Earthquake" in 2015.[6] Its objectives are to help people understand the risks and ways to protect themselves and their community, share preparation information, and emphasize the necessity of doing so. Using an illustrated cartoon of a family of four, it provides basic, easily comprehensible information, with videos and simple graphics, about earthquakes, their risks, ways to prepare before they happen, and the best ways to act when they do. With support from the Romanian Ministry of Research, Innovation, and Digitalization (Ministerul Cercetarii, Inovarii si Digitalizarii, MCID), the INCDFP launched the PreQuake Project (https://prequake.infp.ro/), which hosts links to a wealth of information, including an online interactive map titled "Bucharest's Seismic Risk," a visual representation of buildings assigned to a seismic risk class or urgency level as well as buildings pending demolition.[7] Additionally, it links to or hosts video documentaries on both the 1940 and 1977 earthquakes; scientific documents and books concerning Romanian earthquakes and building codes; an online and in-person visual tour, "Bucharest and Its Earthquakes"; and photos of locations in Bucharest before their damage from an earthquake.[8]

Notes

1 See INCDFP, "Bucureştii şi cutremurele."
2 Administraţia Municipală, "Listă imobile expertizate."
3 Ministerul Afacerilor Interne, "Concepţia natională."
4 Ministerul Afacerilor Interne.
5 INCDFP, "Seismicitate recentă."
6 Inspectoratul General pentru Situaţii de Urgenţa, "Nu tremur la cutremur."
7 For the map, see Pre-Quake Project, "Riscul seismic al Bucureştiului."
8 For the digital tour, see INCDFP, "Bucureştii şi cutremurele."

Bibliography

Archival Materials

Arhivele Naţionale ale României [Romanian National Archives] (ANR)

- Colecţia Consiliul de Stat al P.S.R. Decrete Prezidenţiale [Collection of the State Council of the Romanian Socialist Party, presidential decrees] (CCSPSR).
- Fondul Commitetul Central al Partidul Comunist Român [Central Committee of the Romanian Communist Party Files] (FCCPCR).
    - Secţia Administrativ-Politică (1921–77), inventory 3059.
    - Secţia Cancelarie (1976–77), vol. 6, inventory 3125.
    - Secţia Cancelarie (1978–79), vol. 7, inventory 3129.
    - Secţia Cancelarie (1980–1989), vol. 8, inventory 3354.
    - Secţia Economică (1966–77), vol. 2, inventory 3130.
    - Secţia Economică (1978–89), vol. 3, inventory 3294.
    - Secţia Economică, Dosar Anexe (1950–89), inventory 3148.
    - Secţia Propagandă si Agitaţie (1977–89), inventory 3291.

Arhivele Naţionale ale României şi Institutul de Investigare a Crimelor Comunismului în Romania [Romanian National Archives and Investigative Institute of Communist Crimes in Romania] (ANR and IICCR)

Fototeca online a comunismului românesc [Online photo collection of Romanian communism].

Consiliul Naţional pentru Studierea Arhivelor Securităţii [National Council for the Study of the Securitate Archives] (CNSAS)

- Fond Documentar [Documentary Files] (D)
    - D.123, v. 9. "Referitor la hotariri ale cons. de ministri decrete ale cond. de stat" (1972).
- Fond Documentar Bucureşti [Bucharest Documentary Files] (DB).
    - D.11.487, v. 2. "Referitor şedinţele Biroului Executive al Consiliuli de Conducere al M.I. din 22 februarie şi 19 martie 1977."

- D.11.487, v. 4. "Referitor şedinţele Biroului Executive al Consilui de Conducere al M.I. din 21 mai şi 27 iunie 1977."
- D.11.487, v. 10. "Rapoarte privind îndeplinirea ordinelor, note informative de punere în aplicare a ordinelor primite de la M.I.; plan de măsuri pentru siguranţa circulaţiei scrisorilor" (April–June 1977).
- D.11.737, v. 104. "Buletine informative, note şi situaţii statistice cu privire la cutremurul din 4 martie 1977" (vol. 1).
- D.11.737, v. 105. "Buletine informative, note şi situaţii statistice cu privire la cutremurul din 4 martie 1977" (vol. 2).
- D.12.639, v. 15. "Buletine informative, note de evenimente, etc.—Rapoarte, note, buletine informative, adrese privind urmările cutremurului din martie 1977; acţiunile întreprinse de trupele de securitate, pagubele materiale produse, starea de spirit a populaţiei" (1977).
- D.12.639, v. 16. "Buletine informative, note de evenimente, etc.—Note, adrese, buletine informative, sinteze privind cutremurul din 4 martie 1977; materiale trimise din ţară" (1977).
- D.13.339, v. 37. "Buletine informative şi note de evenimente pe linie de securitate în cadrul acţiunii 'Cutremur'—S.M.B." (1977).
- D.14.800, v. 1. "Documente privind cutremurul din 4 martie 1977, tabele cu victimele cutremurului, organizarea şi componenţa Comandamentului de acţiune al M.I." (1977).
- D.14.800, v. 2. "Documentele intocmite în cadrul direcţiei secretariat-juridice sau primate de aceasta direcţie, în legatura cu seismul din 4 martie 1977" (March 1977).
- D.16.334, v.5. "Centrala Construcţii Industriale Bucureşti (C.C.I.B)" (1977).
- D.21, v. 55. "Problema 'Eterul': rezumate ale emisiunilor transmise de radio 'Europa Liberă'; comentarii transmise în cadrul emisiunii 'Actualitatea Românească'" (1983–89).

- Fond Inventarul Fondului Operaţiuni Valutare Speciale [Inventory of the Special Currency Operations Fund Files] (IFOVS).

  - D.31.160, v. 1. "Telegrame, adrese, note şi liste cu privire la actiunea Solidaritatea—ajutoare primite pentru înlătureare efectelor cutremurul din 4 martie 1977, precum şi alte operaţiuni O.V. specifice. Procese verbale privind distrugerea prin ardere a unor telegrame 7.5.1975–15.02.1979."

- Fond Ministerul de Interne / Direcţia Secretariat-Juridică [Interior Ministry / Legal-Secretariat Files] (MI/DS-J).

  - D.3534, v. 4. "Ordine Originale Circulare Nr. 90394–90478" (1976–1977).
  - D.3634, v. 6. "Ministerul de Interne, Ordine Originale Circulare Nr. 90394–90478" (1976–77).

- D.3635, v. 1. "Ministerul de Interne, Ordine Originale, Nr. 1021–1070" (1977).
- D.3635, v. 2. "Ministerul de Interne, Ordine Originale, Nr. 1071–1131" (1977).

Hoover Institution Library and Archives, Stanford University (HILA)

- Floda, Liviu. Papers.
- Mickelson, Sig. Papers.
- Radio Free Europe/Radio Liberty (RFE/RL). Broadcast records.
- Radio Free Europe/Radio Liberty (RFE/RL). Corporate records.
- Radio Free Europe/Radio Liberty (RFE/RL). Romanian Broadcasting Department records.
- Romanian subject collection.

Open Society Archives at Central European University, Budapest (OSA)
Radio Free Europe/Radio Liberty (RFE/RL). Records (online). https://catalog.osaarchivum.org/catalog/jDenek9n

Nonarchival Materials

ActionAid Agency for International Development. "Earthquake Sitrep." WikiLeaks cable 1977BUCHAR01491_c, dated March 5, 1977a. https://wikileaks.org/plusd/cables/1977BUCHAR01491_c.html

ActionAid Agency for International Development. "Earthquake: Sitrep 4 March 8." WikiLeaks cable 1977BUCHAR01634_c, dated March 8, 1977b. https://wikileaks.org/plusd/cables/1977BUCHAR01634_c.html

Action EUR, Bureau of European and Eurasian Affairs. "ECONADS: Recent Economic Trends in Romania." September 14, 1977a. https://wikileaks.org/plusd/cables/1977STATE220640_c.html

Action EUR, Bureau of European and Eurasian Affairs. "Reported Work Stoppage in Major Bucharest Plant." WikiLeaks cable 1977BUCHAR06998_c, dated September 21, 1977b. https://wikileaks.org/plusd/cables/1977BUCHAR06998_c.html

Action EUR, Bureau of European and Eurasian Affairs. "Shortened Work Week Begins for Some." WikiLeaks cable 1978BUCHAR00813_d, dated February 7, 1978. https://wikileaks.org/plusd/cables/1978BUCHAR00813_d.html

Adevarul. "Fostul şef al pompierilor: După cutremur, multe clădiri au luat foc, apa curentă s-a oprit, aşa că am stins incediile cu dejecţii!." March 4, 2011a. http://adevarul.ro/stiri-locale/bucuresti/fostul-sef-al-pompierilor-dupa-cutremur-multe-1377596.html

Adevarul. "Inginerul care a mutat biserici din calea buldozerelor." February 23, 2011b. https://adevarul.ro/stiri-interne/evenimente/inginerul-care-a-mutat-biserici-din-calea-919297.html

Administraţia Municipală pentru Consolidarea Clădirilor cu Risc Seismic. "Listă imobile expertizate." April 17, 2024. https://amccrs-pmb.ro/lista-imobile-2/

Administraţia Municipală pentru Consolidarea Clădirilor cu Risc Seismic. "Programul Naţional de Consolidare a cladirilor cu Risc Seismic (PNCCRS)." February 13, 2023. https://amccrs-pmb.ro/2023/02/programul

Agenţia Română de Presă (AGERPRES). "March 4, 1977—The 55 Seconds Some Never Forget." March 4, 2014. http://www.agerpres.ro/english/2014/03/04/march-4-1977-the-55-seconds-some-never-forget-14-35-43

Alex, admin. "Marturii de la cutremurul din 4 martie 1977." Cutremur.Net, May 4, 2018. https://forum.cutremur.net/threads/marturii-de-la-cutremurul-din-4-martie-1977.365/

Armaş, Iuliana. "Earthquake Risk Perception in Bucharest, Romania." *Risk Analysis* 26, no. 5 (2006): 1223–34.

Ashley, Carl and Mircea A. Munteanu, eds. "Ceauşescu's Leadership Position, Summary." (August 18, 1977). In *Foreign Relations of the United States, 1977–80*, vol. 20, *Eastern Europe*, document 182. Washington, DC: US Government Printing Office, 2015.

Bachman, Ronald D., ed. *Romania: A Country Study*. Washington, DC: Federal Research Division, Library of Congress, 1989.

Bala, Andrei, and Dragoş Toma-Danila. "The Strong Romanian Earthquakes of 10.11.1940 and 4.03.1977: Lessons Learned and Forgotten?" In *The 1940 Vrancea Earthquake: Issues, Insights, and Lessons Learnt*, edited by Radu Văcăreanu and Constantin Ionescu, 29–34. Cham, Switzerland: Springer International, 2016.

Banu, Florian. "Capitaliştii *avant la lettre*: Securitate şi operaţiunile valutare special din anii '80." *Caietele CNSAS* 4, nos. 1–2 (7–8) (2011): 109–36.

Bălan, Ştefan, Valeriu Cristescu, and Ion Cornea. *Cutremurul de pămînt din România de la 4 martie 1977*. Bucureşti: Editura Academiei Republicii Socialiste Romania, 1982.

Becker, Frederick A. Interview by Charles Stuart Kennedy, 2004. In *Romania Country Reader*, by Association for Diplomatic Studies and Training, 182–96. Accessed March 14, 2024. https://adst.org/Readers/Romania.pdf

Berg, Glen V. "Earthquake in Romania, March 4, 1977: A Preliminary Report." *Earthquake Engineering Research Institute Newsletter* 11, no. 3B (May 1977).

Berg, Glen V., Bruce A. Bolt, Mete A. Sozen, and Christopher Rojahn. *Earthquake in Romania, March 4, 1977: An Engineering Report*. National Research Council and Earthquake Engineering Research Institute. Washington, DC: National Academy Press, 1980.

Bernard, Noel. *Aici e Europa Liberă*. Munich: IDV, 1983.

Betea, Lavinia. *"Tovarăşa": Biografia Elenei Ceauşescu*. Bucureşti: Corint, 2021.

Biel, Steven. *American Disasters*. New York: New York University Press, 2001.

Boia, Lucian. *Strania istorie a comunismului românesc (şi nefericitele ei consecinţe)*. Bucureşti: Humanitas, 2016.

Bottoni, Stefano. "State Violence and Social Control in Communist Romania." *Sources and Methods* (blog). Wilson Center, July 9, 2018. https://www.wilsoncenter.org/blog-post/state-violence-and-social-control-communist-romania

Bucur, Maria. *Heroes and Victims: Remembering War in Twentieth-Century Romania*. Bloomington: Indiana University Press, 2009.

Calotescu, Ileana, Cristian Neagu, and Dan Lungu. "Before and after November 10th, 1940, Earthquake." In *The 1940 Vrancea Earthquake: Issues, Insights, and Lessons Learnt*, edited by Radu Văcăreanu and Constantin Ionescu, 35–56. Cham, Switzerland: Springer International, 2016.

Capotescu, Cristian. "Continuing Politics by Other Means: Giving in Cold War Europe." *International Journal for History, Culture and Modernity* 6 (2018): 105–33.

Capotescu, Cristian. "Migrants into Humanitarians: Ethnic Solidarity and Private Aid-Giving during Romania's Historic Flood of 1970." *East European Politics and Societies* 35, no. 2 (May 2021): 293–312.

Ceauşescu, Nicolae. "Last Speech." December 21, 1989. YouTube videos, uploaded by Avocatul Poporului. Part 1, 9:45, posted April 22, 2010, https://www.youtube.com/watch?v=wWIbCtz_Xwk. Part 2, 9:58, posted August 23, 2019, https://www.youtube.com/watch?v=3muDjmU2ozQ

Ceauşescu, Nicolae. *Report of the National Conference of the Romanian Communist Party, July 19–21, 1972*. Bucharest: Meridiane, 1972.

Chirot, Daniel. "Social Change in Communist Romania." *Social Forces* 57, no. 2 (December 1978): 457–99.
Consiliului de Stat al Republicii Socialiste România privind înfiinţarea, organizarea şi funcţionarea Ministerului de Interne. Decret nr. 130/1972, April 19, 1972. Accessed April 14, 2024. http://www.cnsas.ro/documente/istoria_sec/documente_securitate/xorganizare_interna/1972%20Decret.pdf
Consiliul National pentru Studierea Arhivelor Securităţii (CNSAS). "Ancheta P.C.R. din 1968 privind abuzurile Securităţii." Accessed February 17, 2024a. http://www.cnsas.ro/ancheta_1968.html
Consiliul National pentru Studierea Arhivelor Securităţii (CNSAS). "Index de termini şi abrevieri cu utilizare frecventa în documentele securităţii." Accessed February 18, 2024b. http://www.cnsas.ro/documente/arhiva/Dictionar%20termeni.pdf
Consiliul National pentru Studierea Arhivelor Securităţii (CNSAS). "Ministerul de Interne, 1948–1978." Accessed February 18, 2024c. http://www.cnsas.ro/documente/cadrele_securitatii/Ministerul%20de%20Interne%20-1948-1978.pdf
Danta, Darrick. "Ceauşescu's Bucharest." *Geographical Review* 83, no. 2 (April 1993): 170–82.
Deletant, Dennis. *Ceauşescu and the Securitate: Coercion and Dissent in Romania, 1965–1989*. New York: M. E. Sharpe, 1995.
Despa, Oana. "Starvation, Abuse: New Details Emerge about the Horrors of Romania's Communist-Era Orphanages." Radio Free Europe/Radio Liberty, December 3, 2023. https://www.rferl.org/a/romania-communist-orphanages-starvation-abuse/32711948.html
Direcţia Regională de Statistică a Municipiului Bucureşti. "Populaţia după domiciliu la 1 ianuarie 2022 pe grupe de vârstă şi sexe la nivelul Municipiului Bucureşti şi a sectoarelor administrative." March 2023. https://bucuresti.insse.ro/wp-content/uploads/2023/03/Populatia-dupa-domiciliu-la-1-ianuarie-2022_date-finale.xls
Dobre, Florica, ed. *Membrii C.C. al P.C.R., 1945–1989: Dicţionar*. Consiliul National pentru Studierea Arhivelor Securităţii. Bucureşti: Editura Enciclopedică, 2004.
Dobre, Florica, Florian Banu, Luminiţa Banu, and Laura Stancu. *Acţiunea "Recuperarea": Securitatea şi emigrarea germanilor din România (1962–1989)*. Bucureşti: Editura Enciclopedică, 2001.
Drcimage. "Veneţia din suflete." YouTube video, 11:13, posted March 25, 2012. https://www.youtube.com/watch?v=WUJ2yTSAN1s
Enyedi, György. "Urbanization under Socialism." In *Cities after Socialism: Urban and Regional Change and Conflict in Post-Socialist Societies*, edited by Gregory Andrusz, Michael Harloe, and Ivan Szelenyi, 100–118. Oxford: Blackwell, 1996.
Eskenasy, Victor. "Cazul Gheorghe Ursu şi Radio Europă Liberă." Radio Free Europe/Radio Moldova, November 18, 2015. https://moldova.europalibera.org/a/27373046.html
Gage, Joe. "Memorialul Bucureştilor - Episodul 8 - Casa Poporului." YouTube video, 44:49, posted December 16, 2012. https://www.youtube.com/watch?v=g5P5ylj70Y4
Galmeanu, Alex. "Filmul demolarii bisericii Enei—Aprilie 1977." *Muzeul de Fotografie* (blog), November 19, 2016. https://www.muzeuldefotografie.ro/2016/11/filmul
Gatejel, Luminiţa. "Bargaining for Humanitarian Aid across the Iron Curtain: Western Relief Workers in Romania in the Late 1970s." *Cold War History* 22, no. 1 (April 2021): 41–57.
Georgescu, Diana. "Small Comrades as Historians and Ethnographers: Performativity, Agency, and the Socialist Pedagogy of Citizenship in Ceauşescu's Romania, 1969–1989." *Slavic Review* 78, no. 1 (Spring 2019): 74–102.
Georgescu, Emil-Sever. "The Collapse of Carlton Building in Bucharest at November 10, 1940, Earthquake: An Analysis Based on Recovered Images." In *The 1940*

Vrancea Earthquake: Issues, Insights, and Lessons Learnt, edited by Radu Văcăreanu and Constantin Ionescu, 57–72. Cham, Switzerland: Springer International, 2016.
Georgescu, Emil-Sever. "Earthquake Engineering Development before and after the March 7, 1977, Vrancea, Romania Earthquake." *Geofizica* 1 (2003): 93–107.
Georgescu, Emil-Sever, and Antonios Pomonis. "Building Damage vs. Territorial Casualty Patterns during the Vrancea (Romania) Earthquakes of 1940 and 1977." Paper presented at the Fifteenth World Conference on Earthquake Engineering, Lisbon, September 24–28, 2012. http://www.iitk.ac.in/nicee/wcee/article/WCEE2012_2123.pdf
Georgescu, Emil-Sever, and Antonios Pomonis. "The Romanian Earthquake of March 4, 1977, Revisited: New Insights into Its Territorial, Economic and Social Impacts and Their Bearing on the Preparedness for the Future." Paper presented at the Fourteenth World Conference on Earthquake Engineering, Beijing, October 12–17, 2008. http://www.iitk.ac.in/nicee/wcee/article/14_10-0013.PDF
Georgescu, Emil-Sever, Antonios Pomonis, and Karin Steinbrueck. "New Archival Evidence on the 1977 Vrancea, Romania Earthquake and Its Impact on Disaster Management and Seismic Risk." Paper presented at the Sixth National Conference on Earthquake Engineering and Second National Conference on Earthquake Engineering and Seismology, Bucharest, June 14–17, 2017.
Giurescu, Dinu C. *The Razing of Romania's Past: International Preservation Report*. US Committee of the International Council on Monuments and Sites. Washington, DC: Preservation Press, 1989.
Goma, Paul. *Culoarea curcubeului '77 (Cutremurul oamenilor) / Cod "Barbosul."* Iași, Romania: Polirom, 2005.
Graham, Stephen, ed. *Disrupted Cities: When Infrastructure Fails*. New York: Routledge, 2010.
Hamilton, F. E. Ian. "Urbanization in Socialist Eastern Europe: The Macro-Environment of Internal City Structure." In *The Socialist City: Spatial Structure and Urban Policy*, edited by R. A. French and F. E. Ian Hamilton, 167–94. Chichester, UK: Wiley, 1979.
Hewitt, Kenneth. *Regions of Risk: A Geographical Introduction to Disasters*. Harlow, UK: Longman, 1997.
Hitchins, Keith. *A Concise History of Romania*. Cambridge: Cambridge University Press, 2014.
Hossu Longin, Lucia. *Memorialul durerii: Întuneric și lumină*. Bucureşti: Humanitas, 2013.
Hubbard, Ben, and Safak Timur. "He Warned Turkey about Earthquakes: Now He Fears for Istanbul." *New York Times*, October 6, 2023.
Immigration and Refugee Board of Canada. "Rank Hierarchy of Officers in the Romanian Army and the Securitate [ROM13169]." European Country of Origin Information Network, document 1025186, February 12, 1993. https://www.ecoi.net/en/document/1025186.html
Informaţia Bucureştiului: Ziar al Comitetului Municipal Bucureşti al P.C.R şi al Consiliului Popular al Municipiului Bucureşti. 1954–89.
Inspectoratul General pentru Situaţii de Urgenţa. "Nu tremur la cutremur." Accessed February 25, 2024. https://nutremurlacutremur.ro/
Institutul Naţional de Cercetare Dezvoltare pentru Fizica Pământului (INCDFP). "Bucureştii şi cutremurele." Digital tour, December 2020. https://arcg.is/5PzKe
Institutul Naţional de Cercetare Dezvoltare pentru Fizica Pământului (INCDFP). "Cutremurul vrâncean din 4 martie 1977." YouTube video, uploaded by INCD pentru Fizica Pământului, 32:22, posted March 3, 2022. https://www.youtube.com/watch?v=U_QsdLKcACQ
Institutul Naţional de Cercetare Dezvoltare pentru Fizica Pământului (INCDFP). "Seismicitate recentă." Accessed February 25, 2024. https://infp.ro

Istorie Religie Civilizaţii. "Casa Poporului – Mapa Speciala – Omul si timpul." YouTube video, 10:09, posted April 2, 2012. https://www.youtube.com/watch?v=0gCHhMXm7XI
Janos, Andrew. *East Central Europe in the Modern World: The Politics of Borderlands from Pre- to Postcommunism*. Stanford, CA: Stanford University Press, 2002.
Johnson, A. Ross. *Radio Free Europe and Radio Liberty: The CIA Years and Beyond*. Stanford, CA: Stanford University Press, 2010.
Johnson, A. Ross, and R. Eugene Parta, eds. *Cold War Broadcasting: Impact on the Soviet Union and Eastern Europe*. Budapest: Central European University Press, 2010.
Judt, Tony. *Postwar: A History of Europe since 1945*. New York: Penguin, 2005.
King, Charles. *The Moldovans: Romania, Russia, and the Politics of Culture*. Stanford, CA: Hoover Institution Press, 2000.
Kligman, Gail. *The Politics of Duplicity: Controlling Reproduction in Ceauşescu's Romania*. Berkeley: University of California Press, 1998.
Lambru, Steliu. "Currency Exchange Operations of the Securitate." *Radio Romania International*, February 2, 2013. https://www.rri.ro/en_gb/currency_exchange_operations_of_the_securitate-1342
Leahu, Gheorghe. *Bucureştiul disparut*. Bucureşti: Editura Arta Grafică, 1995.
Livezeanu, Irina. *Cultural Politics in Greater Romania: Regionalism, Nation Building, and Ethnic Struggle, 1918–1930*. Ithaca, NY: Cornell University Press, 1995.
Lynn, Jonathan. "Cigarettes Good as Gold in Romania." *Montreal Gazette*, December 1, 1984.
Machedon, Luminiţa, and Ernie Scoffham. *Romanian Modernism: The Architecture of Bucharest, 1920–1940*. Cambridge: MIT Press, 1999.
Massino, Jill. "From Black Caviar to Blackouts: Gender, Consumption, and Lifestyle in Ceauşescu's Romania." In *Communism Unwrapped: Consumption in Cold War Eastern Europe*, edited by Paulina Bren and Mary Neuberger, 226–54. New York: Oxford University Press, 2012.
Massino, Jill. "Something Old, Something New: Marital Roles and Relations in State Socialist Romania." *Journal of Women's History* 22, no. 1 (Spring 2010): 34–60.
Mauch, Christof, and Christian Pfister, eds. *Natural Disasters, Cultural Responses: Case Studies toward a Global Environmental History*. Lanham, MD: Lexington Books, 2009.
MediaFax. "Mărturii din 4 martie 1977: Erau mormane de moloz şi oameni care scormoneau să scoată oameni." *Ziarul Financiar*, March 3, 2012. http://www.zf.ro/zf-24/focus-marturii-din-4-martie-1977-erau-mormane-de-moloz-si-oameni-care-scormoneau-sa-scoata-oameni-9366440
Mihalache, Cornel, dir. *Babu: Cazul Gheorghe Ursu*. VideoDialog Studio: Bucureşti, 1995. Film, 63 min. Gheorghe Ursu Foundation videotapes, box MC22, Romanian subject collection, Hoover Institution Library and Archives, Stanford University, Stanford, CA.
Ministerul Afacerilor Interne, Departamentul Pentru Situaţii de Urgenţă. "Concepţia natională de răspuns post seim—ediţia a II-a." Inspectoratul General pentru Situaţii de Urgenţă, 2021. https://www.igsu.ro/Resources/biblioteca/Instructiuni/Conceptie%20Nationala%20de%20Raspuns%20Post%20Seism.pdf
Nagat, Germina. "Ceauşescu's War against Our Ears." In *Cold War Broadcasting: Impact on the Soviet Union and Eastern Europe*, edited by A. Ross Johnson and R. Eugene Parta, 229–38. Budapest: Central European University Press, 2010.
Neagoe-Pleşa, Elis. "1968—Anul reformării agenturii Securităţii." *Caietele CNSAS* 1, no. 1 (2008): 9–22.
Neagoe-Pleşa, Elis, and Liviu Pleşa. *Securitatea: Structuri / Cadre: Obiective şi metode*. Vol. 2, *1967–1989*. Edited by Florica Dobre. Bucureşti: Editura Enciclopedică, 2006.

New York Times. "Hundreds Reported Killed by Earthquake in Rumania." March 5, 1977, 1.

Noica, Nicolae. "DOSAR 77/45: Ce am facut, dar mai ales ce putem face cu cladirile degradate?" *Revista Construcţiilor* 17, no. 181 (June 2021): 8–10.

Partidul Comunist Român. "25 iunie 1984 Inaugurarea oficială a lucrărilor de construcţie a Casei Republicii." YouTube video, 2:32, posted January 18, 2015. https://www.youtube.com/watch?v=cE9MsOLkkx0

Partidul Comunist Român. "Cutremurul din Bucureşti 4 martie 1977." YouTube video, 2:31, posted January 25, 2015. https://www.youtube.com/watch?v=xeF1WwGe2c0

Petrescu, Cristina. "The Afterlife of the Securitate: On Moral Correctness in Postcommunist Romania." In *Remembering Communism: Private and Public Recollections of Lived Experience in Southeast Europe*, edited by Maria Todorova, Augusta Dimou, and Stefan Troebst, 385–415. Budapest: Central European University Press, 2014.

Petrovici, Radu. "Architecture Face to Face with Past and Future Earthquakes." Paper presented at Simpozionul Naţional SPAŢIU, ARTĂ, ARHITECTURĂ, Bucharest, April 20–21, 2013.

Petrovici, Radu. "The 10 November 1940—The First Moment of Truth for Modern Constructions in Romania." In *The 1940 Vrancea Earthquake: Issues, Insights, and Lessons Learnt*, edited by Radu Văcăreanu and Constantin Ionescu, 84–99. Cham, Switzerland: Springer International, 2016.

Pittaway, Mark. *Eastern Europe, 1939–2000*. New York: Bloomsbury, 2004.

Popa, Cosmin. *Elena Ceauşescu, sau, Anatomia unei dictaturi de familie*. Bucureşti: Litera, 2021.

Popa, Raluca Maria. "Understanding the Urban Past: The Transformation of Bucharest in the Late Socialist Period." In *Testimonies of the City: Identity, Community and Change in a Contemporary Urban World*, edited by Richard Rodger and Joanna Herbert, 159–86. Aldershot, UK: Ashgate, 2007.

Pre-Quake Project. "Riscul seismic al Bucureştiului" / "The Seismic Risk of Bucharest." December 2022. https://prequake.infp.ro/

Pucci, Molly. *Security Empire: The Secret Police in Communist Eastern Europe*. New Haven, CT: Yale University Press, 2020.

Raab, Nigel. *All Shook Up: The Shifting Soviet Response to Catastrophes, 1917–1991*. Montreal: McGill-Queen's University Press, 2017.

Radio Free Europe/Radio Liberty. "Romania, 1989." Accessed February 17, 2024. https://pressroom.rferl.org/p/7341.html (page discontinued, accessible through the Internet Archive).

Răduţă, Cristina. "Cinematograful Patria a fost închis pentru că funcţiona într-o clădire cu bulină roşie." *Adevarul*, December 4, 2015. https://adevarul.ro/stiri-locale/bucuresti/cinematograful-patria-a-fost-inchis-pentru-ca-1671612.html

Re: Rise, Asociaţia pentru Reducerea Riscului Seismic. "Bucureştul după cutremur." Accessed February 17, 2024. https://sites.google.com/rerise.org/dupacutremur/home

Ronnås, Per. "Centrally Planned Urbanization: The Case of Romania." *Geografiska Annaler: Series B, Human Geography* 64, no. 2 (1982): 143–51.

Roper, Steven D. *Romania: The Unfinished Revolution*. London: Routledge, 2000.

Rosca, Matei. "IKEA Funds Went to Romanian Secret Police in Communist Era." *Guardian*, July 4, 2014. Accessed February 17, 2024. https://www.theguardian.com/world/2014/jul/04/ikea-funds-romania-secret-police-communist-era

Sampson, Steven L. "Urbanization—Planned and Unplanned: A Case Study of Brasov, Romania." In *The Socialist City, Spatial Structure and Urban Policy*, edited by R.A. French and F.E. Ian Hamilton, 507–525. Chichester: John Wiley and Sons, 1979.

Sampson, Steven L. *National Integration through Socialist Planning*. Boulder, CO: East European Monographs, 1984.

Shafir, Michael. *Romania. Politics, Economics, and Society: Political Stagnation and Simulated Change*. London: Frances Pinter, 1985.
Siani-Davies, Peter. *The Romanian Revolution of December 1989*. Ithaca, NY: Cornell University Press, 2007.
Solnit, Rebecca. *A Paradise Built in Hell: The Extraordinary Communities That Arise in Disaster*. New York: Viking, 2009.
Stan, Lavinia. "Inside the Securitate Archives." Cold War International History Project. Wilson Center, March 4, 2005. https://www.wilsoncenter.org/article/inside-the-securitate-archives
Thurow, Roger. "In Romania, Smoking a Kent Cigarette Is Like Burning Money: The U.S.-Made Brand Reigns as Prime Barter Medium; Sure Sign of a Big Shot." *Wall Street Journal*, January 3, 1986.
Tismăneanu, Vladimir. *Stalinism for All Seasons: A Political History of Romanian Communism*. Berkeley: University of California Press, 2003.
Tismăneanu, Vladimir. *The Tragicomedy of Romanian Communism*. Research report for National Council for Soviet and East European Studies. Foreign Policy Research Institute, September 1989. https://www.ucis.pitt.edu/nceeer/1989-903-04-2-Tismaneanu.pdf
Tismăneanu, Vladimir, Dorin Dobrincu, and Cristian Vasile, eds. *Comisia Prezidenţială pentru Analiza Dictaturii Comuniste din România: Raport final*. Bucureşti, 2006. Wilson Center. https://www.wilsoncenter.org/sites/default/files/media/documents/article/RAPORT%20FINAL_%20CADCR.pdf
Tofan, Liviu, ed. *Antologia Radio Europa Liberă: 101 Contribuţii (1956–1990)*. Bucureşti: Editura Omnium, 2021.
Toma-Danila, Dragoş, and Iuliana Armaş. "Insights into the Possible Seismic Damage of Residential Buildings in Bucharest, Romania, at Neighborhood Resolution." *Bulletin of Earthquake Engineering* 15, no. 3 (2017): 1161–84.
Toma-Danila, Dragoş, Alexandru Tiganescu, Mircea Radulian, and Dragoş Tataru. "Earthquakes in Romania: Lessons of the Past, Brought to the Present." Paper presented at the Third European Conference on Earthquake Engineering and Seismology, Bucharest, September 4–9, 2022.
Transparency International. "Romania Country Data." Accessed February 14, 2024. https://www.transparency.org/en/countries/romania
Tulbure, Narcis. "Drink, Leisure, and the Second Economy in Socialist Romania." In *Pleasures in Socialism: Leisure and Luxury in the Eastern Bloc*, edited by David Crowley and Susan E. Reid, 259–82. Evanston, IL: Northwestern University Press, 2010.
Turnock, David. "The Planning of Rural Settlement in Romania." *Geographical Journal* 157, no. 3 (November 1991): 251–64.
Ujica, Andrei, dir. *The Autobiography of Nicolae Ceauşescu*. Bucureşti: ICON Production, 2010. Film, 186 min.
Universitatea Tehnică de Construcţii Bucureşti. *Thirty Years from the Romania Earthquake of 4 March 1977*. Bucharest: Romanian National Center for Earthquake Engineering and Vibrations, 2007.
Ursu, Gheorghe. *Europa Mea*. Cluj-Napoca, Romania: Editura Dacia, 1991.
US Congress. "Romanian Earthquake Relief," September 8, 1977. In *Foreign Assistance and Related Agencies Appropriations for 1978: Hearings before a Subcommittee of the Committee on Appropriations, House of Representatives*, 95th Cong., 1st sess., part 4, 33–67. Washington, DC: US Government Printing Office, 1977.
US Geological Survey. "Earthquake Early Warning around the World." *Earthquake Hazards Program*, March 9, 2022. https://www.usgs.gov/programs/earthquake-hazards/science/earthquake-early-warning-around-world

Vale, Lawrence J., and Thomas J. Campanella, eds. *The Resilient City: How Modern Cities Recover from Disaster*. New York: Oxford University Press, 2005.

Vatulescu, Cristina. *Police Aesthetics: Literature, Film, and the Secret Police in Soviet Times*. Stanford, CA: Stanford University Press, 2010.

Verdery, Katherine. *National Ideology under Socialism: Identity and Cultural Politics in Ceauşescu's Romania*. Berkeley: University of California Press, 1995.

Verdery, Katherine. "Romania's Securitate Archive and its Fictions: An Introduction." Working paper, National Council for Eurasian and East European Research, June 24, 2013. https://www.ucis.pitt.edu/nceeer/2013_826-01g_Verdery.pdf

Verdery, Katherine. *Secrets and Truths: Ethnography in the Archive of Romania's Secret Police*. Budapest: Central European University Press, 2014.

Verdery, Katherine. *What Was Socialism, and What Comes Next?* Princeton, NJ: Princeton University Press, 1996.

Vlad, Ion. "Causes and Effects of the November 10, 1940, Earthquake." In *The 1940 Vrancea Earthquake: Issues, Insights, and Lessons Learnt*, edited by Radu Văcăreanu and Constantin Ionescu, 113–28. Cham, Switzerland: Springer International, 2016.

Vlaicu, Ion. "În orele grele şi lungi ale durerii, cadrele aparatului de securitate s-au aflat în prima linie a eroicului front al încrederii şi reconstrucţiei!" *Securitatea* 5, no. 1 (37) (1977): 9–94.

Vultur, Smaranda. "Daily Life and Surveillance in the 1970s and 1980s." In *Remembering Communism: Private and Public Recollections of Lived Experience in Southeast Europe*, edited by Maria Todorova, Augusta Dimou, and Stefan Troebst, 417–36. Budapest: Central European University Press, 2014.

Washington Post. "Transcript of the Closed Trial of Nicolae and Elena Ceausescu." December 29, 1989.

White Mountain Property. "Bucharest 1977: The Last Major Earthquake." YouTube video, 5:02, posted December 6, 2016. https://www.youtube.com/watch?v=3rdsgzF9PLg

World Bank. *Report and Recommendation of the President of the International Bank for Reconstruction and Development to the Executive Directors on a Proposed Loan to the Investment Bank with the Guarantee of the Socialist Republic of Romania for a Post Earthquake Construction Assistance Project*. Report no. P-2240-RO, May 17, 1978.

Index

Pages in *italics* refer to figures and pages followed by 'n' refer to notes.

For Product Safety Concerns and Information please contact our EU representative GPSR@taylorandfrancis.com
Taylor & Francis Verlag GmbH, Kaufingerstraße 24, 80331 München, Germany

www.ingramcontent.com/pod-product-compliance
Ingram Content Group UK Ltd.
Pitfield, Milton Keynes, MK11 3LW, UK
UKHW020004200926
475702UK00015B/2467
* 9 7 8 1 0 3 2 6 3 2 9 2 6 *